T0280550

INTRODUCTION TO AIRCRAFT DESIGN

This book provides an accessible introduction to the fundamentals of civil and military aircraft design.

Giving a largely descriptive overview of all aspects of the design process, this well-illustrated account provides an insight into the requirements of each specialist in an aircraft design team. After discussing the need for new designs, the text assesses the merits of different aircraft shapes from micro-lights, UAVs and helicopters to super-jumbos and V/STOL aircraft. Following chapters explore structures, airframe systems, avionics and weapons systems. Later chapters examine the costs involved in the acquisition and operation of new aircraft, aircraft reliability and maintainability, and a variety of past aircraft projects to see what conclusions can be drawn. Three appendices and a bibliography give a wealth of useful information, much not published elsewhere, including simple aerodynamic formulae, aircraft, engine and equipment data and a detailed description of a parametric study of a 500-seat transport aircraft.

Introduction to Aircraft Design is a useful text for undergraduate and graduate aeronautical engineering students and a valuable reference for professionals working in the aerospace industry. It should also be of interest to aviation enthusiasts.

Professor J. P. Fielding, M.Sc., Ph.D., CEng., FRAeS, SMAIAA had 12 years industrial experience as an apprentice and design engineer on the Nimrod, BAE SYSTEMS 748 and BAE SYSTEMS 146 aircraft. He then joined Cranfield University as a research fellow, with subsequent promotions to professor and Head of the Department of Aerospace Engineering. He was also Chief Engineer of the BAE SYSTEMS/EPSRC–sponsored Demon technology demonstrator UAV. He has partially retired and is now Professor Emeritus at Cranfield University. He is founder and director of Fielding Aerospace Consultants Ltd. He specialises in research and education in aircraft initial project design, reliability, maintainability and operational effectiveness. He acts as a private consultant and expert witness in the fields of aircraft design, maintainability and reliability and has published more than 100 technical papers at conferences and in technical journals. Recent presentations were made at the ICAS 2014 Conference in St. Petersburg, Russia and ICAS 2016 in Korea. He is a member of the international advisory boards of a large Chinese aircraft company and of a leading aero engine manufacturer.

CAMBRIDGE AEROSPACE SERIES

Editors: Wei Shyy and Vigor Yang

1. J. M. ROLFE and K. J. STAPLES (eds.): *Flight Simulation*
2. P. BERLIN: *The Geostationary Applications Satellite*
3. M. J. T. SMITH: *Aircraft Noise*
4. N. X. VINH: *Flight Mechanics of High-Performance Aircraft*
5. W. A. MAIR and D. L. BIRDSALL: *Aircraft Performance*
6. M. J. ABZUG and E. E. LARRABEE: *Airplane Stability and Control*
7. M. J. SIDI: *Spacecraft Dynamics and Control*
8. J. D. ANDERSON: *A History of Aerodynamics*
9. A. M. CRUISE, J. A. BOWLES, C. V. GOODALL, and T. J. PATRICK: *Principles of Space Instrument Design*
10. G. A. KHOURY (ed.): *Airship Technology*, Second Edition
11. J. P. FIELDING: *Introduction to Aircraft Design*
12. J. G. LEISHMAN: *Principles of Helicopter Aerodynamics*, Second Edition
13. J. KATZ and A. PLOTKIN: *Low-Speed Aerodynamics*, Second Edition
14. M. J. ABZUG and E. E. LARRABEE: *Airplane Stability and Control: A History of the Technologies that Made Aviation Possible*, Second Edition
15. D. H. HODGES and G. A. PIERCE: *Introduction to Structural Dynamics and Aeroelasticity*, Second Edition
16. W. FEHSE: *Automatic Rendezvous and Docking of Spacecraft*
17. R. D. FLACK: *Fundamentals of Jet Propulsion with Applications*
18. E. A. BASKHARONE: *Principles of Turbomachinery in Air-Breathing Engines*
19. D. D. KNIGHT: *Numerical Methods for High-Speed Flows*
20. C. A. WAGNER, T. HÜTTL, and P. SAGAUT (eds.): *Large-Eddy Simulation for Acoustics*
21. D. D. JOSEPH, T. FUNADA, and J. WANG: *Potential Flows of Viscous and Viscoelastic Fluids*
22. W. SHYY, Y. LIAN, H. LIU, J. TANG, and D. VIIERU: *Aerodynamics of Low Reynolds Number Flyers*
23. J. H. SALEH: *Analyses for Durability and System Design Lifetime*
24. B. K. DONALDSON: *Analysis of Aircraft Structures*, Second Edition
25. C. SEGAL: *The Scramjet Engine: Processes and Characteristics*
26. J. F. DOYLE: *Guided Explorations of the Mechanics of Solids and Structures*
27. A. K. KUNDU: *Aircraft Design*
28. M. I. FRISWELL, J. E. T. PENNY, S. D. GARVEY, and A. W. LEES: *Dynamics of Rotating Machines*
29. B. A. CONWAY (ed): *Spacecraft Trajectory Optimization*
30. R. J. ADRIAN and J. WESTERWEEL: *Particle Image Velocimetry*
31. G. A. FLANDRO, H. M. MCMAHON, and R. L. ROACH: *Basic Aerodynamics*
32. H. BABINSKY and J. K. HARVEY: *Shock Wave–Boundary-Layer Interactions*
33. C. K. W. TAM: *Computational Aeroacoustics: A Wave Number Approach*
34. A. FILIPPONE: *Advanced Aircraft Flight Performance*
35. I. CHOPRA and J. SIROHI: *Smart Structures Theory*

INTRODUCTION TO AIRCRAFT DESIGN

JOHN P. FIELDING
Cranfield University

CAMBRIDGE
UNIVERSITY PRESS

CAMBRIDGE
UNIVERSITY PRESS

University Printing House, Cambridge CB2 8BS, United Kingdom

One Liberty Plaza, 20th Floor, New York, NY 10006, USA

477 Williamstown Road, Port Melbourne, VIC 3207, Australia

314-321, 3rd Floor, Plot 3, Splendor Forum, Jasola District Centre, New Delhi - 110025, India

79 Anson Road, #06-04/06, Singapore 079906

Cambridge University Press is part of the University of Cambridge.

It furthers the University's mission by disseminating knowledge in the pursuit of
education, learning and research at the highest international levels of excellence.

www.cambridge.org
Information on this title: www.cambridge.org/9781107680791

First published 1999
Second edition 2017

A catalogue record for this publication is available from the British Library

Library of Congress Cataloging in Publication data
Names: Fielding, John P., 1945- author.
Title: Introduction to aircraft design / John P. Fielding, Cranfield University.
Description: Second edition. | New York, NY : Cambridge University Press, 2017. |
Includes bibliographical references and index.
Identifiers: LCCN 2016030834 | ISBN 9781107680791 (pbk.)
Subjects: LCSH: Airplanes–Design and construction.
Classification: LCC TL671.2 .F46 2017 | DDC 629.133/34–dc23 LC record
available at https://lccn.loc.gov/2016030834

ISBN 978-1-107-68079-1 Paperback

DEDICATION

This book is dedicated to the late Professor David Keith-Lucas, CBE, and Professor Denis Howe, both of whom were the author's immediate predecessors as Professor of Aircraft Design at the College of Aeronautics, Cranfield University, The author gained much of his knowledge of aircraft design, much encouragement and good role-models from these elder-statesman of aircraft design education, as well as strong support from the late Professor John Stollery, CBE.

CONTENTS

PREFACE

This book acts as an introduction to the full breadth of both civil and military aircraft design. It is designed for use by senior undergraduate and post-graduate aeronautical students, aerospace professionals and technically inclined aviation enthusiasts.

The book poses and answers pertinent questions about aircraft design, and in doing so gives information and advice about the whole aircraft design environment. It asks why we should design a new aircraft and gives examples of market surveys and aircraft specifications. It then answers the question, 'Why is it that shape?' and gives the rationale behind the configurations of a wide range of aircraft from micro-lights and helicopters to super-jumbos and V/STOL aircraft, with many others in between. Having looked at the shape, the book then examines and describes what is under the skin in terms of structure, propulsion, systems and weapons. Later chapters answer questions about aircraft costs and conceptual design and draw lessons from past projects, and then look into the future. A major part of the book answers the question, 'What help can I get?', with a combination of bibliography, lists of data sheets, computer tools and 100 pages of appendices of design data vital to aircraft conceptual designers (most of it previously unpublished).

The book concentrates on fixed-wing civil and military aircraft, with some reference to light aircraft and rotorcraft, but does not address the design of sailplanes, airships, flying boats or spacecraft. While these are fascinating and important subjects it was decided that the current scope of the book is sufficiently wide and further extension would make it unwieldy, although information about references that address the design of aircraft in the excluded categories is provided.

Much of the material has been developed for use in Pre-Masters and Masters' courses in aircraft design at Cranfield University. Many of the examples and illustrations have been produced as part of Cranfield's unique group design project programmes. With the Cranfield method, conceptual design is done by the staff, thus enabling the students to start much further down the design process. They thus have the opportunity to get to grips with preliminary and detail design problems, and become much more employable in the process. This method also allows students to use modern design tools such as CAD, finite elements, laminate analysis and aerodynamic modelling. The group design project is undertaken by all the aerospace vehicle design students and is a major feature of the M.Sc. course, accounting for almost half of the final assessment. Each year the students work in a team on the design of a project aircraft. A substantial part of the airframe, a system, an installation or some performance aspect is allocated to each student with his or her own responsibility. Most students also undertake a secondary task in areas of project management, CAD integration, mass and centre of gravity (CG) control, and the construction of aircraft physical models. The aircraft chosen as the subject for the work are representative of types of current interest to industry. They usually incorporate some feature that extends the bounds of existing practice, as an applied research activity. This excites the interest, enthusiasm and ingenuity of the students and forces the staff to keep up to date. Civil and military aircraft, and spacecraft are investigated, so that the whole of the industry is catered for. Recent examples of design projects included large and small business jets, a number of medium- and large-sized, environmentally friendly jet transports and a 500-seat short-haul airliner. The latter aircraft is described in Chapter 10 and Appendix B of this book. Military aircraft have also been designed, including basic and advanced trainers, close-air support aircraft, an advanced tactical fighter, V/STOL supersonic strike aircraft, UASs and military transports.

Cranfield University has developed a part-time M.Sc. in Aircraft Engineering. The group design project is, again, at the heart of this programme, with a duration of three years. This allows the design process to be extended into detail design, manufacture and, in some cases, a prototype flight test. A good example is the Kestrel blended wing body unmanned flying demonstrator.

There are many textbooks available that cover the conceptual design phase and others that provide the more specific information appropriate to the detail design phase. This book has been written to fill the gap between these two stages, utilizing the experience gained from all the projects carried out at Cranfield and from other industrial projects.

ACKNOWLEDGEMENTS

The author would like to acknowledge the assistance of current and past students and staff of Cranfield University. Thanks are particularly due to Professor Howard Smith, the author's successor as Professor of Aircraft Design at Cranfield University. He was responsible for the conceptual design of most of the recent Cranfield Group Design Project aircraft.

The book contains more than 200 illustrations and large amounts of data. Much of this has come from the author's and colleagues' work at Cranfield University, from whom permission has been received for publication.   The author would also like to acknowledge the help of the following individuals, companies and organizations that have given publishing permission, and provided data, drawings or photographs:

Airbus: Figs. 2.1 3.4, 3.5, 3.14, 5.13, 6.12, 8.1, A9.2, A9.3, A9.8, A9.11, A9.12, A9.13.

American Society of Mechanical Engineers: Fig. 4.5.

Boeing Commercial Airplane Company: Figs. 2.2, 2.3, 2.4, 3.6, 3.15, 8.6.

Bombardier Commercial Aircraft: Fig. 3.10

BAE SYSTEMS: Figs. 4.23, 4.25, 6.3, 6.4, 6.5, 6.6, 6.7, 6.8, 6.9, 6.13, 7.2, 7.3, 8.19, 8.20, 8.21, 11.1, 11.6, 11.15, 11.24.

A British airline: Figs. 8.15, 8.16, C2.1.

Davey, Bernard, Air Cargo Research Team: Fig. 11.19.

Denel Ltd South Africa: Fig. 4.14.

Doganis, R Ref.17: Fig. 8.6

Dunlop Aerospace Ltd – data used in Tables A8.1 and A8.2.

EMBRAER, Brazil: Figs. 3.9, 8.15.

European Space Agency: Fig. 11.12.

Fairchild Aircraft Company: Figs. 4.9, 6.11.

Flight International Magazine – some of the data used in Tables A4.1–A4.11.

Fokker Aircraft: Figs. 5.5, A.2.1, A9.6, A9.7, A9.9, A9.10.

Jane's All the World's Aircraft: some of the data used in Tables A4.1–A4.11.

Ministry of Defence, UK: Figs. 8.11, A9.1 and A9.4.

Northrop Ltd: Fig. 11.8.

SAAB Aircraft Company: Figs. 3.11 and A9.5.

Salamander Books: Fig. 11.4.

Solar Wings: Figs. 3.20.

Norman Wijker: Fig. 1.3.

CHAPTER 1   INTRODUCTION

1.1 WHY ANOTHER AIRCRAFT DESIGN BOOK?

Aircraft design is a complex and fascinating business and many books have been written about it. The very complexity and dynamic nature of the subject means that no one book can do it justice.

This book, therefore, will primarily act as an introduction to the whole field of aircraft design leading towards the subjects summarized in Fig. 1.1. It will not attempt to duplicate material found in existing design books, but will give information about the whole aircraft design environment, together with descriptions of aircraft and component design. It also presents otherwise unpublished data and design methods that are suitable for aircraft conceptual, preliminary and detail design activities.

1.2 TOPICS

The following chapters are arranged as a series of questions about aircraft design, the answers to which give largely descriptive overviews of all aspects of aircraft design. This will provide an introduction to the conflicting requirements of aircraft design specialists in a design team, with a view to improving understanding and the integration of a sound overall design.

The book is divided into chapters that answer a number of significant design questions.

The question, 'Why design a new aircraft?' is answered in Chapter 2, which shows the derivation of aircraft requirements for civil and military aircraft from market surveys, and gives examples of operator-derived specifications.

Chapters 3 and 4 answer the question, 'Why is it that shape?' with an initial discussion of aircraft wing and tail shapes, followed by descriptions of the configurations of a wide range of civil and military aircraft types.

The question, 'What's under the skin?' is answered in Chapters 5, 6 and 7, which deal with structures and propulsion, airframe systems, avionics, flight controls and weapons, respectively. These chapters describe the interiors of aircraft, ranging from structures to weapons systems via airframe systems, avionic systems and landing gears.

In Chapter 8 the crucial areas of acquisition and operating costs are discussed, some prediction methods are described and the importance of good reliability and maintainability are stressed in order to answer the question, 'Why do aircraft cost so much?'

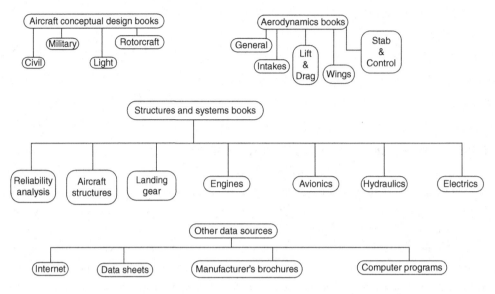

Fig. 1.1 Aircraft design data sources.

The answer to the question, 'What help can I get?' is provided in Chapter 9, which contains a bibliography of the most important current aircraft design books. It is followed by a description of some of the computer design analysis and computer-aided design (CAD) tools that are available. A summary of relevant data sheets is also given.

Chapter 10 draws together the information produced at the end of the conceptual stage and leads on to the preliminary and detail design stages in order to explain, 'What happens next?'. The question, 'What can go wrong?' is answered in Chapter 11, in which many unsuccessful or partially successful projects are examined and conclusions drawn from them.

The aircraft designer is bedeviled by lack of design data. Appendix A pulls together information that is not generally available, and includes simple aerodynamic and structural design formulae. It also provides a US/British translation list for aeronautical terms.

Appendix B presents a parametric study design example that describes the author's parametric study of a 500-seat transport aircraft. Appendix C considers reliability and maintainability targets by discussing targets for civil and military aircraft, and describing a method to be used for the prediction of dispatch reliability.

1.3 THE DESIGN PROCESS

There are a number of generally accepted stages in the design, development, manufacture and operation of aircraft, each with associated design methods and data requirements. These are shown schematically in Fig. 1.2, which also shows how the modern practice of concurrent engineering has reduced the overall timescale from conception to service.

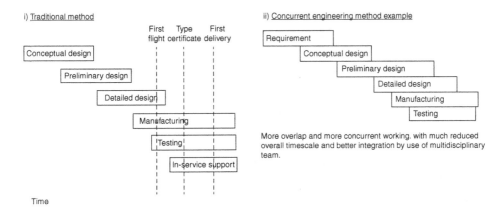

i) Traditional method

First flight | Type certificate | First delivery

- Conceptual design
- Preliminary design
- Detailed design
- Manufacturing
- Testing
- In-service support

Time

ii) Concurrent engineering method example

- Requirement
- Conceptual design
- Preliminary design
- Detailed design
- Manufacturing
- Testing

More overlap and more concurrent working. with much reduced overall timescale and better integration by use of multidisciplinary team.

Fig. 1.2 Comparison of traditional and concurrent design approaches.

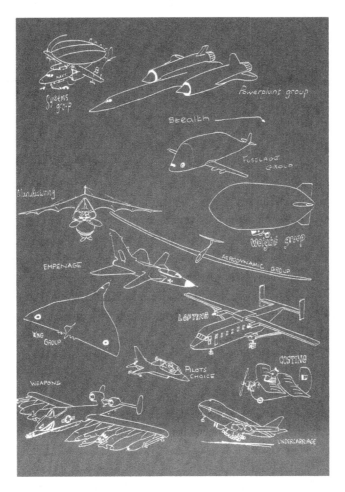

Fig. 1.3 Different specialist's views of an ideal aircraft.

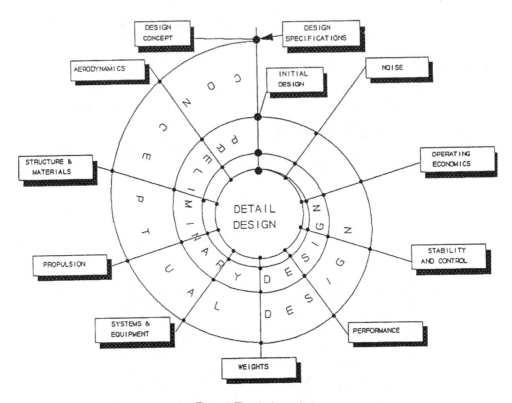

Fig. 1.4 The design spiral.

Figure 1.3 gives some idea of how a designer's prejudice may affect his or her design to the detriment of others. It is an exaggeration, but not much of an exaggeration!!

The most crucial stage of any design process is to arrive at the correct set of requirements for the aircraft. These are summarized in design specifications for the particular aircraft type. Typical examples of design specifications are shown in Chapter 2. They are augmented by a large number of airworthiness requirements for civil aircraft or defence standards for military aircraft. These are distillations of decades of successful (and unsuccessful!) design, manufacturing and operational experience. Figure 1.4, adapted from Haberland et al. [1], shows a very helpful illustration of what may happen after the issue of the design specifications, and illustrates the iterative design process that is not apparent in the simplified illustration in Fig. 1.2.

A converging iterative spiral of design stages, ending in the detailed design, and ultimately manufacture and operation of the aircraft, can be seen in Fig. 1.4.

It is a truism that 99% of the decisions that affect aircraft success are made on 1% of the facts available during the conceptual design phase. Very coarse methods have to be used, which are then refined by progressively more accurate methods as the design evolves. This is true if the spiral

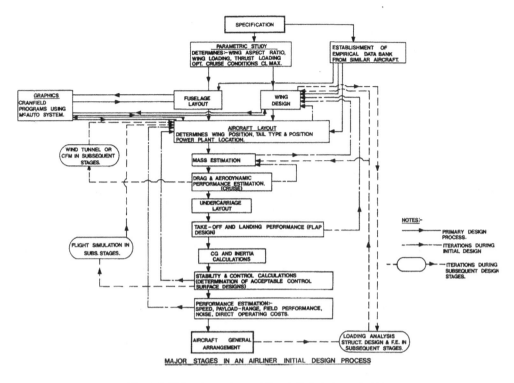

Fig. 1.5 Major stages in an airliner initial design process.

is convergent, but there are occasions where the spiral is divergent and the design must be abandoned, and started again, unless significant modifications are made to the design.

Figure 1.5 shows the author's usual design procedure for conceptual design and the start of the preliminary design process.

CHAPTER 2

WHY SHOULD WE DESIGN
A NEW AIRCRAFT?

The world has accepted that flying is an extremely efficient means of quickly transporting people, cargo or equipment, and performing a wide range of other activities. All operators need to increase efficiency, cost-effectiveness, environmental compatibility and safety. It is often possible to do this by modifying either the design or operation of existing aircraft. This is limited, however, by the inherent capabilities of the original design and the cost-effectiveness of modifications. Under these circumstances, it is necessary to consider the initiation of the design of a new aircraft. Aircraft manufacturers are usually in the business of making profit out of building aircraft. They may do this by means of building their own existing designs, modifying their designs or licence-building the designs of other companies. Another reason for the initiation of new designs is to retain or enhance their design capabilities.

Safety is the most important aspect of aircraft design. This is the moral and legal responsibility of designers, manufacturers and operators of both civil and military aircraft. The regulations that have emerged aim to maximize safety, and must be complied with before an aircraft may receive a Type Certificate to allow it to be sold. A brief summary of such requirements is contained in Chapter 9.

In recent years there has been a considerable increase in the importance of requirements associated with aircraft-related effects on the environment. These particularly concern noise, airport local air quality and global warming. Current environmental requirements are summarized in Chapter 9.

Operator requirements and aircraft specifications come from a number of different sources, but they must all consider the needs of the aircraft operators, whether they are airlines or air forces. A certain path to disaster is to produce an aircraft that no one will buy!

Descriptions will be given of the two main means of deriving a requirement specification, namely the results of market surveys and individual aircraft operators' specifications.

2.1 MARKET SURVEYS

The major aircraft manufacturers employ marketing departments that produce annual reports [2]. Historical data are analyzed and extrapolated in such areas as world economic indicators. Figure 2.1 shows gross domestic product (GDP) and highlights the recent recessions, and shows the close correlation between GDP and the world air travel growth in revenue passenger miles (RPM).

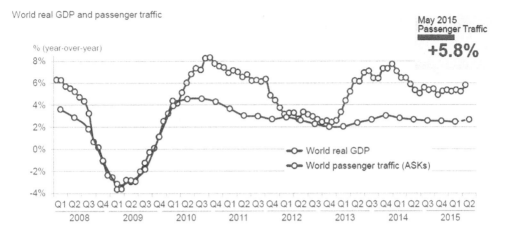

Fig. 2.1 Recent changes in world gross domestic product (GDP) 2015 [3].

Fig. 2.2 Recent changes in world revenue passenger miles (RPM), 2015 [2].

Further work led to predictions of world capacity requirements in terms of passenger aircraft required (Fig. 2.2) and numbers of passenger seats in various aircraft size categories (Fig. 2.3). Figure 2.4 shows an estimate of the commercial aircraft global market Table 2.1 shows the results of the author's summary of the market surveys which led to the specification for the aircraft design described [5]. Appendix B shows the initial stages of a commercial aircraft design process.

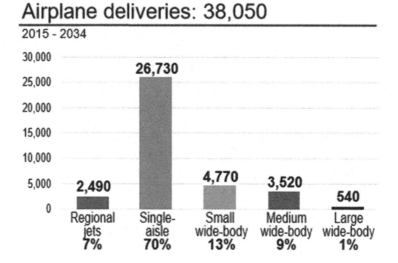

Fig. 2.3 Commercial aircraft capacity projections 2013 [2].

Having determined the projected market for aircraft, the next stage was to examine existing and proposed competitors in that market. Figure 2.5 shows a very simple but effective illustration of current and projected jet transport competitors. More details of these aircraft are gathered and analyzed to show strengths and weaknesses. Some useful means of comparison are the tables of aircraft major characteristics, as shown in Appendix A, as part of what might be the rather grandly named 'empirical database'. This sort of data is an obvious candidate for computerization. An annual publication is the major source of the information required for such tables [6], but it may be augmented by information from aeronautical journals and manufacturers' brochures. General arrangement drawings such as shown in Fig. 2.6 yield valuable information.

The final stage in the derivation of the specification is to determine targets of how much to improve certain performance parameters, comfort or costs, relative to the competition. The aircraft must be significantly better in important areas to stand a chance of selling well, unless there is political pressure to buy a certain aircraft. Payload range diagrams of competing aircraft are very instructive (Fig. 2.7), as are comparative direct operating costs. The magnitude of the specified improvements are a source of judgment and are risky. Too great an improvement may not be achievable and too little may be too small to be attractive. This can be checked to some extent during the conceptual design stage and the specification may be modified.

2.2 OPERATOR-DERIVED SPECIFICATIONS

Military aircraft specifications are often derived using operational research techniques, but include significant input from pilots and engineers. They rely heavily on data from existing aircraft that

Table 2.1 Summary of future requirements for airliners (2015)

Parameter	Boeing [2]	Airbus [3]
Projected date	2034	2034
Projected fleet	38050	32600
Percentage of 400 + passengers	1	5
Percentage of 250 + passengers	9	25
Percentage of small twin aisle	13	
Percentage of single aisle	70	70

Fig. 2.4 Current aircraft market [2].

need to be replaced or to be augmented by other aircraft. The perceived threat is an important element of the specification and is determined from intelligence sources, with more than a small amount of 'crystal-ball gazing'.

More complex aircraft are often procured by several governments and follow the production of multi-national specifications. Airlines or groups of airlines know the capabilities of their current fleets and make predictions of traffic on current and future routes. Short-falls lead to specifications for new aircraft for their *particular* requirements. The aircraft designer has to be careful not to follow individual airline requirements to the exclusion of other potential customers. There have been sad cases when this has happened and later competitors have been more flexible and much more successful.

The following paragraphs are modified extracts from the ground attack mission of a fighter aircraft specification and show what are termed the cardinal points, and also show those for an uninhabited combat air vehicle (UCAV).

Further paragraphs show the major points of a specification from a large US airline. This specification was written in imperial units, which are usual in the Western air transport industry.

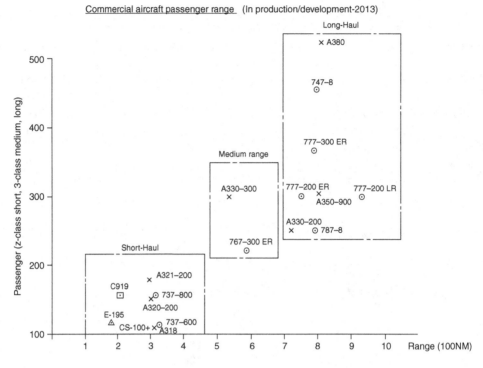

Fig. 2.5 Airliner seat/range capabilities, 2014.

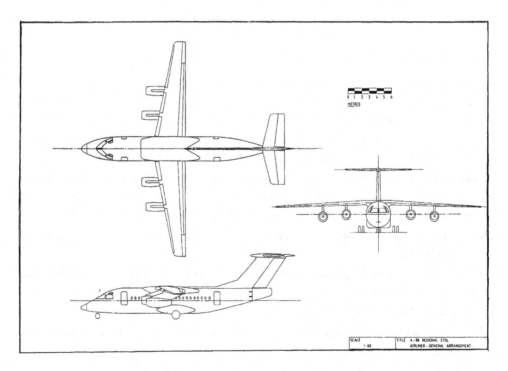

Fig. 2.6 Cranfield regional jet general arrangement drawing, 1988.

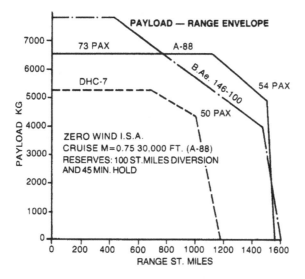

Fig. 2.7 Regional aircraft payload-range envelopes.

2.3 SPECIFICATION FOR A CLOSE AIR SUPPORT AIRCRAFT

2.3.1 Modes of Operation

The need exists for a close air support capable of operating in support of ground troops in the vicinity of the forward edge of battle area (FEBA). Three modes of operation are required.

2.3.1.1 Autonomous Mode
In this mode the aircraft is required to operate some 100 km behind the FEBA. The targets will be first and second echelon troops – in particular main battle tanks, armoured fighting vehicles, armoured personnel carriers and their support vehicles in daylight hours.

2.3.1.2 Forward Air Control Mode
In this mode, the aircraft is directed on to targets by a forward air control aircraft (FAC). It is envisaged that the FAC aircraft would carry the complex electronic systems required to direct, identify and mark targets.

2.3.1.3 Ground Control Mode
In this mode the aircraft is directed onto targets by ground forces.

2.3.2 Mission Radius

The radius of action of the aircraft should be 500 km at 3.5 km altitude, optimum speed with a 15 min loiter/search time and a 10% fuel reserve. These figures are based upon the following mission at the specification payload:

- (i) Cruise at optimum speed at low level 300 km to FEBA.
- (ii) Accelerate to Mach Number (M) 0.8 at a height of 100 m and carry out mission. (100 km penetration to target.)
- (iii) Optimum speed cruise at low level 300 km back to base.

2.3.3 Take-off and Landing Performance

The aircraft should take off from semi-prepared strips in a distance of 880 m over a 15 m obstacle. Landing ground distance should be 500 m.

2.3.4 Performance

The following performance figures are required:

- (i) Maximum speed: 0.8 at sea level (SL) with a specified payload (2.3.5 below); 0.9 at SL with two advanced short-range air-to-air missiles (ASRAAM) alone.
- (ii) Maximum operational height: 3500 m.
- (iii) Maximum-sustained g loading: $4g$ at specified payload at M 0.8 at SL.

2.3.5 Weapon Loads

Possible weapon loads include:

- (i) Six 277 kg cluster bombs + two ASRAAM + gun + ammunition (specified payload).
- (ii) Eight 500 kg low drag bombs + two ASRAAM + gun.
- (iii) One JP233 fuselage store of 2335 kg + two ASRAAM + gun.
- (iv) Six short range air to ground guided missiles + two ASRAAM + gun.

2.3.6 Avionics fit

The avionics fit comprises the following systems:

Integrated communication, navigation and identification system.
Inertial navigation system.

Controls and displays.

Radar altimeter.

Digital computers.

Weapon management and interface unit (IFU).

Tail warning (0.15 m diameter antenna).

Radar warning receiver.

2–18 GHz jammer.

Chaff and flare dispensers.

Ground attack, laser range finder.

Forward-looking infra-red (FLIR).

2.4 REQUIREMENTS FOR AN UNINHABITED COMBAT AIR VEHICLE (UCAV)

Figure 2.8 shows the initial requirements for a new class of aircraft, the UCAV, [7]. It was derived from a specification used in an annual AIAA student design competition. This, in turn was arrived at by experienced aircraft design engineers from industry, government employees and academics. Figure 2.9 shows a CAD image of the Cranfield University U-99 group design project aircraft, designed to meet the above specification. Note that the air intakes are not the final shape.

The aircraft is defined by the requirement implied by a primary mission, secondary mission and ferry mission. These mission imply weapons carriage and other operational constraints. Point performance and airlift constraints are also be satisfied

Design missions
High altitude deep penetration (primary mission)
-Cruise segment 1 (outbound leg) ⁓ 750 nm
     @ 40,000 ft
     @ best mach
-Weapon release (454 kg) from appropriate altitude and speed with
1.5 turns.
-Cruise segment 2 (inbound leg) as cruise segment 1
-6% fuel remaining

Hi-Lo-Hi strike (secondary mission)
-Cruise segment 1 (out bound leg high altitude)
Range as dictated by fuel required for primary mission
     @ 40,000 ft
     @ best mach
-Cruise segment 2 (out bound leg low altitude)
     100 nm ingress
     @ 250 ft
     @ mach 0.9
-Weapon release (454 kg) with 1.5 turns
-Cruise segment 3 (in bound leg low altitude)
     100 nm egress
     @ 250 ft
     @ mach 0.9
-Climb
-Cruise segment 4 (Inbound leg high altitude)
     @ 40,000 ft
     @ best mach
-6% fuel remaining
Ferry mission
     3000 nm
     @ best speed and altitude
     Auxiliary fuel tanks may be
used (e.g. in weapons bay)

Field performance
     Take-off & 3500 ft
     Landing & 3500 ft

Fig. 2.8 UCAV Requirements, 1999.

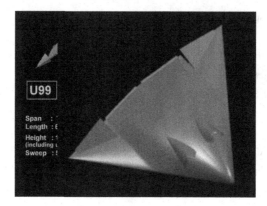

Fig. 2.9 Cranfield U-99 UCAV project.

2.5 AIRLINE SPECIFICATION FOR A 150-SEAT AIRLINER

This is an amalgamation of US and European specifications, that were current in mid-1980s, and led to the MD80 and Airbus A320-type designs.

2.5.1 Introduction

The aircraft is to be a 150-passenger, short-range, twin-engined aircraft designed and constructed to the latest technology in both airframe and propulsion systems. The major design objectives are to include an aircraft with minimum seat mile cost and yet provide maximum comfort to the passenger. It is expected that the design will utilize the latest aerodynamic, systems and material technology consistent with the stated mission requirements. At the same time, design demands will be extended to achieve a maximum service life, a maximum of reliability and full ease of maintainability. A minimum requirement for training, both ground and flight, is to be a goal. Spares requirements are to be minimized and maximum utilization of present ground equipment will be paramount. Fuel conservation will be a major consideration and all present and pending environmental requirements must be fully considered in the design.

Each design iteration shall be fully optimized to the criteria without considering further growth or derivative aircraft of the type.

2.5.2 Capacity Requirements

The interior configuration will consist of the following specified items.

2.5.2.1 Interior

(i) Capacity – A total of 150 passengers: 138 tourist class and 12 first class.

(ii) Seats – Pitch 36" in first class, 32" in tourist. Lightweight bulkhead seats to have built-in armrest tray tables. Width between armrests: first class 19.5" with 4" outer and 8" centre armrests; tourist class 17.5" minimum with 2" armrest.

(iii) Aisle width – 18" minimum with adequate spacing at galley and coatrooms to allow ease of access and service.

(iv) Attendants' seats – A total of five (5) seats shall be installed. Restraints and cushioning will be designed for maximum service utility.

(v) Flight deck – Design shall include convertibility to either two or three man crew.

(vi) Galley area – Space for one full hot meal service for 150 passengers.

(vii) Lavatory – Design will include at least three (3). Floor plan layout shall include tradeoffs for four (4).

(viii) Overwater provisions – Shall include slide rafts or life rafts and life-vest stowage.

(ix) Passenger air – Each passenger shall have individual air (gasper) outlets.

2.5.2.2 Cargo

(i) Doors – Locations shall allow maximum bin space utilization and access without a mechanized loading system.

(ii) Ventilation and temperature – Adequate to allow carriage of live animals in designated areas.

2.5.2.3 General External Dimension Criteria

The outlined design characteristics generally dictate external dimensions of the aircraft to be:

Span	110–120 ft
Length	120–135 ft
Tail Height	30–45 ft

2.5.3 Operating Profile

2.5.3.1 Trip Length

An operating fleet average of 370 statute miles. The distribution of individual flight segments by distance and present allocation to trip length is shown in Table 2.2 and a model of the Cranfield A-82 that was designed to meet this specification is shown in Fig. 2.10.

Table 2.2   Typical flight and trip distance profile

Segment distance (statute miles)	0–150	150–300	300–450	450–600	600–750	750–900	900–1050	1050–1200
Percentage of total flight in segment	15	26	25	14	7	5.5	5.0	2.5

Fig. 2.10 Cranfield A82 short-range jet transport, 1984.

2.5.3.2 Payload and Range

The minimum trip length with a full space limit payload shall be 1000 statute miles (SM). A passenger and bag payload range shall be on the order of 1400 SM miles.

2.5.4 Operational Characteristics

2.5.4.1 Flight

(i) Speeds – The maximum design cruise speed will be of the order of M 0.83; minimum cost cruise from M 0.76 to M 0.78. Maximum speed on approach will be 130 knots.

(ii) Altitude – The maximum cruise altitude shall be of the order of 39000 with a minimum initial cruise altitude of 31000 feet after a maximum weight take-off.

A single engine approved performance enroute altitude shall be 16000 feet (ISA + 10 °C) at the midpoint of 1200 SM full passenger flight.

(iii) Fuel efficiency – The objectives in ASM per gallon while carrying a full passenger and baggage payload shall be:

Trip length (SM)	ASM/ gallon
400	77.6
600	85.0
1000	92.0

2.5.4.2 Ground

(i) Airport compatibility: runway – The maximum certificated runway required for a trip of 370 SM with a space limit payload shall be 6000 ft on a 90 °F day at SL. The pavement and flotation design criteria shall preclude limitations upon the aircraft at any standard commercially acceptable airport.

(ii) Noise: exterior – Noise levels shall be lower than FAR 36 Stage 3 limits without requiring trades or thrust reductions for compliance. Levels shall meet all required standards for ground service personnel.

(iii) Emissions – Engine emissions will meet all projected standards and design effort will be expended to achieve levels well below standards.

(iv) Wheel and engine placement – Wheel and engine placement design shall be such as to attempt to preclude foreign object damage (FOD). Design criteria in engine placement, air flow and ground clearance for all ground operation, including takeoff or reverse thrust conditions shall contain detail to minimize FOD.

(v) Ground times – Service access, equipment and door design shall be configured to achieve the maximum ground times of: (a) through flights, 15 min; (b) turnaround 30 min.

2.5.5 Reliability and Maintainability

(i) Design for mechanical dispatch reliability shall include a full system analysis of the highest order with a goal of achieving a 99.5% mechanical reliability by the end of first year of operation.

(ii) The most advanced cost-effective materials shall be utilized to minimize weight and drag and in all instances, particular attention shall be centred upon corrosion prevention, protection and/or treatment.

(iii) All aircraft systems shall have prime consideration for maintainability in design and placement. Unit removal and replacement times will be reduced to a minimum and full demonstration of access, elapsed removal and replacement time and tools will be conducted.

(iv) Overall design fatigue life shall fully consider the short cycle time flight.

2.5.6 Design Considerations

2.5.6.1 Aerodynamics

The latest technology in aerodynamics shall be utilized. Specifically:

(i) Aerofoils and control surfaces contour and geometry shall use advanced technology, optimized for the intended mission.

(ii) Active controls shall be used to minimize weight and drag. Full consideration should be given to both load alleviation and stability augmentation type systems (wing and tail).

2.5.6.2 Structural

(i) Brake and wheel systems shall be adequate to handle short cycles separately without affecting minimum ground times. Full consideration should be given to the use of carbon brakes.

(ii) All pressure vessel doors shall be plug type.

2.5.6.3 Systems

(i) Air conditioning – Environmental control system (ECS) shall be air cycle cooling, with percentage recirculation to be decided during trade-off studies.

(ii) Hydraulics – Hydraulic power generation shall place minimum reliance on air-driven and electric hydraulic pumps. The use of screw-jack type actuators should be minimized.

(iii) Automatic flight – The automatic flight control system shall make extensive use of digital avionics. The system shall contain an automatic subsystem with fail active CAT IIIb certification. A fully integrated flight management system(s) shall be considered.

(iv) Landing gear – Nosewheel steering shall be capable of steering angles of $\pm 90°$ nominal with torque link connected.

(v) Flight controls – Primary consideration shall be given to single-segment trailing-edge flaps. Control surface sealing shall be easily replaceable.

(vi) Communications – Digital to include provisions for triple VHF and single ADF.

(vii) Navigation – digital – Standard very-high-frequency omnidirectional range (VOR), instrument landing system (ILS), with provisions for microwave landing system (MLS).

(viii) Auxiliary power – Trade studies should be generated on the requirement for an airborne use of an auxiliary power unit (APU).

(ix) Instruments – Consideration shall be given to electronic display types.

2.5.6.4 *Engines*

(i) Thrust – Takeoff thrust shall be in the 20 000–23 000 lb thrust class flat-rated at least 85°FatSL.

(ii) Weight – Bare engine weight shall be in the 4000–4300 lb range as a design goal.

(iii) Life – Rotating components shall be designed for a 30 000 cycle service life. A shop visit rate of 0.35 per 1000 cycles shall be a design goal.

(iv) Deterioration – 1% fuel consumption deterioration for the guarantee shop visit rate period and 2% long-term deterioration shall be design goals. Case distortion, transient rubbing, and blade/stator erosion shall be minimized through optimum combinations of materials selection and active clearance controls.

(v) Modularization – Modules and mini-modules shall be designed for both ease of disassembly/assembly and minimal post-assembly test cell running. Mating seals between modules shall be designed to prevent both assembly problems and component performance matching problems.

CHAPTER 3 WHY IS IT THAT SHAPE?: CIVIL AIRCRAFT

3.1 BACKGROUND

There is almost no limit to the variety of the shapes of aircraft that have been conceived. This chapter will give a brief introduction describing wing, fuselage and powerplant arrangements, followed by descriptions of the characteristics of civil, military and rotorcraft types.

Aircraft are essentially aerodynamic vehicles and every aircraft designer must develop considerable aerodynamic skills. This book is not aimed at providing those, as there are many excellent texts available, as shown in the bibliography section of this book. Appendix A6 contains some limited information to enable the evaluation of simple aerodynamic predictions for use during the conceptual design process.

The aerodynamic shape of the aircraft will determine aircraft speeds, manoeuvrability, flying qualities, range, field performance, costs, altitudes and many other parameters. There is a close interaction between such disciplines as aerodynamics, structures, propulsion and aircraft systems. For example, a high subsonic aircraft usually requires a very thin or swept wing, or a combination of these, to achieve low aerodynamic drag. Thin or swept wings adversely affect weight and may require more thrust from the propulsion system. A good compromise is therefore required. An optimum shape must be determined to maximize a chosen performance parameter. This may be minimum life-cycle cost, initial cost, weight, fuel burn, etc. depending upon the type of aircraft required. In all cases, however, a good aerodynamic shape is vital, as in the areas now described.

3.1.1 Wings and Tail

The wing is the main lift source and it must be carefully designed to optimize its lift/drag ratio over the whole flight envelope.

Figure 3.1 shows definitions of the main wing geometrical characteristics. The parameters are defined as:

b wing span – maximum distance over wing perpendicular to centreline

s $b/2$, wing semi-span

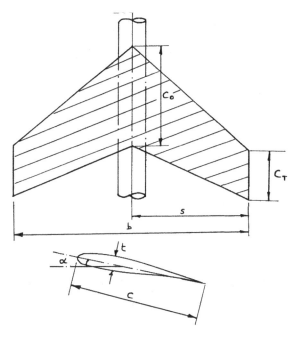

Fig. 3.1 Major wing geometrical definitions.

C_0 root chord of wing defined either as distance at centreline between projections of leading and trailing edges *or* along body side

C_T tip chord of wing defined as the distance between the projection of leading and trailing edges at the tip, measured parallel to the centreline

S wing plan area defined as *gross* area when the area of the body between the lines joining the leading and trailing edges is included and exposed or nett area if this is excluded

λ C_T/C_0, wing taper ratio

t/c wing thickness chord ratio, defined as maximum local thickness divided by local chord length

$\bar{c}$ S/b, geometric wing chord

A b^2/S, wing aspect ratio

α geometric incidence

Δ_{LE} sweep-back (or forward) of the wing leading edge

$\Delta_{1/4}$ sweep-back (or forward) of a line joining the local 25% chord points of a wing.

Some of the more important aerodynamic parameters are:-

L/D Lift/drag ratio, either of wing or complete aircraft, sometimes expressed as C_L/C_D

$$C_L = L/\frac{1}{2}\rho V^2 S = L/\frac{1}{2}\gamma p M^2 S$$

where ρ = air density; p = air pressure; V = velocity; M = Mach number; γ = ratio of specific heats = 1.41 for air;

L = lift; D = drag and $V = Ma_0$, where a_0 = velocity of sound

$$C_D = D / \frac{1}{2}\rho V^2 S = D / \frac{1}{2}\gamma p M^2 S$$

$$C_M = M / \frac{1}{2}\rho V^2 S\bar{c} = D / \frac{1}{2}\gamma p M^2 S\bar{c}$$

where M is pitching moment (nose up positive)

C_{M0} pitching moment coefficient when overall wing or wing-body lift is zero

M_{DR} Mach number where the drag rises significantly

T/W thrust/weight ratio. It is often quoted as powerplant sea level static thrust/ aircraft gross weight. It will, in fact, vary throughout the flight as thrust varies with throttle setting, speed, altitude and temperature. Weight changes with fuel burn and changes of payload

W/S wing loading. The usual parameter is the aircraft gross weight divided by the gross wing area, but again, weight and therefore wing loading will vary through the flight

Camber   The curvature of the aerofoil section. A camber line is defined as the position at mean local thickness relative to the chord line

a_1 Slope of a wing lift curve

V_s Stall speed. This is the condition such that airflow separates and the lift force is reduced, whilst the drag increases rapidly. It occurs immediately after maximum lift $C_{L_{max}}$ and is defined by a re-statement of the lift coefficient equation

$$V_S = \sqrt{\frac{W/S}{\frac{1}{2}\rho C_{L_{max}}}}$$

The stall speed may be reduced by use of leading or trailing edge devices. These increase camber and/or area and thus $C_{L_{max}}$.

Appendix A6 gives simple methods to allow the prediction of lift and drag of aircraft with clean wing (cruise configuration). Wings must be designed to be as efficient as possible over a wide range of often conflicting flight conditions. Table 3.1 highlights the main wing aerodynamic variables, and their effects on different categories of aerodynamic performance.

Propeller-driven aircraft are not likely to be used at Mach numbers in excess of 0.65 and wing sweepback is not therefore necessary. Figure 3.2 shows a typical configuration. The operating lift coefficient can be higher than 0.35 for an 18% thick wing before the critical Mach number falls below 0.65.

Cruise efficiency indicates the need for high aspect ratio coupled with moderate taper of the order of 0.5 to 0.65 to give a close approximation to an elliptical load distribution, which is the best shape for reducing the lift-dependent drag. The tapered wing is easier to manufacture than an

Table 3.1   Wing design features and their effect on performance

Performance requirement	Sweep	Aspect ratio	Thickness of chord	Area	Camber
Stall speed	−	+	+	+	+
Rate of climb	+	+	−	+	−
Absolute ceiling	+	+	−	+	−
Take-off and landing	−	+	+	+	+
Low altitude high subsonic speed	+	0	−	−	−
Gust ride quality	+	−	0	−	0
High Mach no. at altitude	+	−	−	−	−
Supersonic manoeuvre	+	−	−	+	−
Subsonic manoeuvre	−	+	+	+	+

+ = increase in this parameter improves performance.
− = increase in this parameter degrades performance.
0 = increase in this parameter has negligible effect.

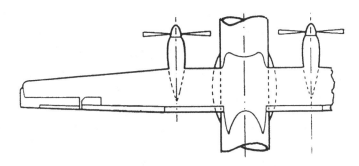

Fig. 3.2 Typical turbo-prop aircraft wing.

elliptical wing and is therefore a better compromise in terms of drag and cost. The upper bound on aspect ratio is fixed by structural weight conditions and is primarily a function of the thickness chord ratio of the aerofoil section, the aircraft manoeuvre factor and the design diving speed.

Typical figures are aspect ratios from 8 to 12 with maximum L/D values of 15 to 18. Root thickness chord ratios usually vary from 14 to 20% and tip values are often 0.65 to 0.75 of the root value.

The wing area has to be chosen as a compromise between that required for cruise and for low speed operation. The use of flaps enables satisfactory low speed performance to be achieved with near optimum cruise wing area. Take-off wing loadings usually vary between 75 and 90 lb/ft^2 (350–450 kg/m^2) and cruise lift coefficients are in the range of 0.3 to 0.45. With lift curve slopes of

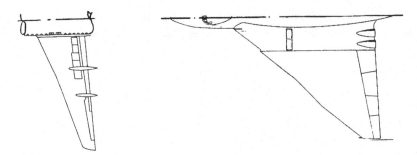

Fig. 3.3 Swept and delta wing planforms.

the order of 6 per radian, this represents only 3–4° of wing incidence and this is normally a fixed setting on the body. The maximum lift coefficient of this type of design is usually around 2.5–3.0 and is achieved by using large chord flaps over about two-thirds of the span. It is frequently sufficient to use the simpler types of flap, in which case the lower value of lift coefficient is appropriate.

As the efficiency of a jet engine increases with Mach number, it is desirable to design the wing of *jet aircraft* such that the value of M_{DR} is as high as is reasonably possible. This is usually in the range $0.78 < M < 0.92$ and attempts to increase it further are probably not worthwhile. The use of sweep is immediately implied for useful lift coefficients (Fig. 3.3). Sweepback is normally used, since swept forward wings are subject to divergence and longitudinal instability unless the angle is small. Sweepback implies higher bending material weight. The thickness chord ratio is usually relatively low, being of the order of 10–15% at the root (the higher with supercritical sections – see Appendix A6 for an explanation of supercritical sections). Stiffness is frequently the design criterion for such wings.

The sweepback required is usually between 15° and 40° and this is associated with moderate taper. Extreme taper can cause tip flow problems, especially at high incidence when 'pitch up' may occur. Pitch up is caused by separation and loss of lift occurring on the outer part of the wing, the remaining lift being inboard and hence well forward of the normal position.

If the wing is mass-balanced by the powerplants, as in several Boeing designs, the aspect ratio can be as high as 10, with root thickness chord ratios of around 15%. Mass-balancing is the term used to describe the wing configuration where the gravity loads of components in or on a wing, such as engines, help to balance the upward lift load on the wing, and reduce wing bending moments. Transport aircraft wings have ample span for flaps and their effectiveness is often improved by reducing the local trailing edge sweep in the root region. Wing area is fixed by cruise considerations and the take-off wing loading may be as high as 150 lb/ft^2 (750 kg/m^2) for large aircraft. Lift curve slope is of the order of 4 to 5 per radian and cruise lift coefficient around 0.3–0.6. Taper ratio is usually about 0.3.

The major advantage gained by using a *delta* planform is the considerable root depth of the wing, even with low thickness chord ratios. There is no difficulty in achieving a high critical Mach number, but by implication the aspect ratio is low. Maximum lift coefficient is low, not exceeding unity in most cases, and because of the relatively small span, flaps do not help a great deal. Wing loading must therefore be low, probably in the region of 45 to 70 lb/ft^2 (200–350 kg/m^2) at take-off, and low speed requirements dictate the choice of wing area. Modification of the wing leading edge shape enables an elliptical load distribution to be approached, and tip flow problems to be mitigated. There is a serious longitudinal stability and control problem which results from the short elevator arm, whether a tailplane is used or not. Since the lift curve slope is low, around 2 or 3 per radian due to the low aspect ratio, the aircraft flies at high incidence at low speed. This gives rise both to vision difficulties for the pilot on the approach and undesirable roll-yaw couplings, which frequently cause instability.

A more recent wing concept combines the advantages of both delta and swept wings. It is often termed the *blended wing body* (BWB). The inner wing is, effectively, a delta wing with a large root chord acting like a fuselage to contain payloads for military aircraft or passengers and cargo for airliners. Examples of BWBs will be shown later in this book.

While the above comments have been applied to wings, they also apply to other flying surfaces such as tailplanes, fins or foreplanes. In American usage these surfaces are called horizontal stabilizer, vertical stabilizer and canard, respectively.

3.1.2 Fuselages

Aircraft fuselage shapes are usually defined by what they have to contain in terms of payload, crew, systems and equipment. Different classes of aircraft have widely varying requirements in terms of payload and their fuselages are consequently of different shapes and sizes. These effects will be discussed below for different classes of aircraft.

3.1.3 Powerplant Choice

The choice of powerplant type and number has a fundamental impact on an aircraft configuration. The usual rule is that the fewer the engines the better, consistent with adequate levels of safety. Fewer engines have simpler aircraft systems and are generally more cost effective in terms of mass, initial and maintenance costs. The four-engined transport aircraft, however, may be optimum for either short-take-off or extremely long-range aircraft. In these cases take-off engine failure requirements with two- or three-engined aircraft may lead to too much power for efficient cruise. Chapter 5 gives the reasoning behind the choice of type and locations of different classes of powerplants.

3.2 CIVIL AIRCRAFT TYPES

3.2.1 Passenger Aircraft

The over-riding requirement for all fare-paying passenger-carrying aircraft is that of safety. The order of safety required is specified by the relevant airworthiness requirements. The two most commonly used requirements are the US Federal Aviation Regulations (FAR) and European Certification Standards (CS). The usual parts in both cases are Part 25 for larger aircraft and Part 23 for aircraft with maximum take-off masses less than 5700 kg (12 500 lb).

Second only to the need for safety is that of economy, since all civil aircraft are operated by organizations which, at least nominally, are profit-making. Aircraft operating costs are divided into two sections. These are basically those directly connected with flying the aircraft, the direct costs, and those resulting from the rest of the organization, the indirect costs. The latter are of little direct concern to the design and cover such items as airport facilities and staff, sales and service, etc.

The direct operating costs (DOC) are discussed more fully in Chapter 8, but fall into the following categories:

   (i) Crew costs.
   (ii) Fuel, oil and taxes.
   (iii) Insurance.
   (iv) Maintenance,which may be approximately divided equally between powerplant maintenance man-hours, powerplant spares, airframe maintenance man-hours and airframe spares.
   (v) Depreciation, which spreads the cost of buying the aircraft. The first cost is usually written off over a 15–20-year operating period and may include interest payments. Table 3.2 summarizes recent airline operating cost figures for a range of current subsonic jet transport aircraft, operated by a large North American airline. The two important parameters are the direct operating cost (DOC) per block hour, that is the time between the aircraft leaving and arriving at passenger terminals. The other measure is the DOC per ASM. Table 3.2 shows direct operating costs for the financial year 2012. Fuel prices are linked to the cost of crude oil, which has recently fluctuated wildly. For example, a barrel of crude oil cost some 100 USD in 2012, for which the data in Table 3.2 applies. In the Spring of 2016, crude oil cost $35! This has a significant effect on aircraft direct operating costs, and the viablilty of future fuel-saving technologies or aircraft configurations. It is likely, however, that fuel prices will increase in the longer term, and potential environmental taxes will encourage the design of more fuel-efficient aircraft.

In general, a large aircraft is cheaper to operate in terms of DOC per ASM than a small one, providing certain conditions are satisfied. This is because the large aircraft has a large payload, but

Table 3.2   Direct operating costs of a large North American airline for the financial year 2012

	Boeing 737-100/200	Airbus A320	Airbus A300	MD-11	Boeing 747–400
Crew (%)	28.8	25.2	14.5	20.9	19.7
Fuel/oil/taxes (%)	25.1	23.5	26.5	32.5	31.3
Insurance (%)	1.6	2.8	1.6	1.4	1
Total flying (%)	55.6	51.5	42.7	54.9	52
Total maintenance (%)	23.9	12.8	28.2	14.4	17.2
Depreciation and rent (%)	20.6	35.7	29.1	30.7	30.8
Total DOC (US$ per block hour)	1702	1808	3644	4399	6592
Number of seats	111	148	262	255	397
Average flights (miles)	441	995	1157	3487	4639
Block hours/day	8.95	10.9	9.2	12.08	13.43
Average load factor (%)	63.1	65.9	67.6	67.4	74.2
Fuel burn US gal/block hour	813	788	1754	2300	3393
Cost/aircraft mile ($)	5.37	4.72	8.25	9.05	12.98
Cost/available seat miles US cents (DOC/ASM)	4.84	3.19	3.54	3.55	3.27

crew, engineering and depreciation costs can be relatively lower. Thus a larger aircraft can operate economically with a smaller percentage payload than a smaller one, that is, it has a lower 'break even' load factor. However it still requires more passengers or freight to reach this load factor and one of the serious problems at present facing airlines is the surplus of capacity over traffic. Thus the general rule is to use the largest aircraft that the traffic route will support. In certain circumstances, however, this may lead to unacceptably low flight frequencies. This will reduce passenger appeal and therefore trade-offs must be performed. Another important fact is that different airlines have different requirements in aircraft range.

All of these factors mean that there is a wide range of successful aircraft types, such as those shown in Table 3.2. They are all efficient in meeting the requirements for their *own* markets, but competitive pressures are leading to the reduction of DOC on future aircraft. Figure 3.4 shows some new aircraft features. Crew costs are reduced by increasingly sophisticated avionics to allow two-crew operation. Many aerodynamic system and structural improvements have led to mass, drag and fuel cost improvements, as have improvements in engine designs and maintainability.

A designer must produce an aerodynamically efficient aircraft equipped with quiet, low-fuel-consumption powerplants. The choice of speed will depend upon this, as well as the design stage length. Payload capacity will depend upon the estimated available traffic and size of operator, but the aircraft must be versatile in being able to trade payload for range, and vice-versa.

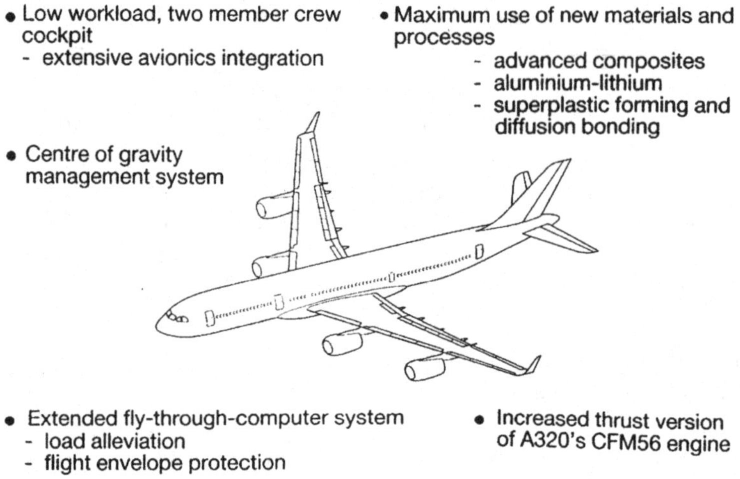

- Low workload, two member crew
  cockpit
  - extensive avionics integration

- Maximum use of new materials and
  processes
  - advanced composites
  - aluminium-lithium
  - superplastic forming and
    diffusion bonding

- Centre of gravity
  management system

- Extended fly-through-computer system
  - load alleviation
  - flight envelope protection

- Increased thrust version
  of A320's CFM56 engine

Fig. 3.4 Some of the advanced design features of the airbus A340 aircraft, 1995.

Long-range airliners have recently polarized into two categories:

Long-range, medium size, which operate point-to-point to a large number of destinations. These are typified by the Boeing 787 Dreamliner (Fig. 3.6) or the more recent Airbus A350 XWB.

Long-range, very large airliners, which operate between a relatively small number of large "hub" airports. The Airbus A380 is a good example of this type (Fig. 3.7)

Shorter-range airliners are designed to cruise most efficiently at Mach numbers between 0.75 and 0.8, whereas longer-range aircraft cruise around M 0.86 at altitudes of around 40 000 ft. The most obvious effect on the aircraft shape is the wing configuration. Earlier jet aircraft, such as the DC-8 and Boeing 747, utilized what were termed high-speed aerofoil sections, which required higher sweep-back than current designs. Quarter-chord sweep angles now vary between 25° and 30°, as against 35°–38°. These reductions are primarily due to the introduction of so-called 'supercritical' aerofoil sections and more sophisticated three-dimensional (3-D) wing-body design techniques, which simultaneously predict the airflow over the junction between the body and wing. These have been made possible by the use of computational fluid dynamic (CFD) programs, allied to the continuing use of wind-tunnels. Reduced sweep improves the wing structural efficiency and makes leading- and trailing-edge flaps more efficient.

The fuselage size is almost entirely determined by the volume required for the maximum number of passengers and maximum quantity of freight, both of which may be selected in various combinations to give the maximum payload. Consistent with meeting this volume requirement, the fuselage must be as small as possible to reduce cruising drag to a minimum. A further compromise is necessary in the design of the wing, since the cruise conditions invariably require a smaller area than those at landing and take-off. The use of flaps may partially overcome this difficulty.

Most airliner manufacturers aim to produce a 'family' of aircraft to cover a wide range of payload-range options.

The Boeing narrow-body family does this by using identical fuselage cross-sections in its 707, 720, 727, 737 and 757 aircraft. This considerably reduces manufacturing costs and gives improved commonality in service. Another method of 'tailoring' aircraft to the market is to stretch the basic aircraft. This is done by the use of parallel section fuselage which is lengthened by inserting 'plugs' of fuselage ahead and behind the wing. This increases the capacity significantly for relatively little extra cost. Most manufacturers use this method, but McDonnell-Douglas have produced the most extreme stretches in their MD80 and 90 ranges.

This process can also be reversed for what are termed 'thin' routes. In these there is not enough traffic to fill, say, a 747 at the required frequencies so fuselage sections are removed. This was done by reducing the Boeing 747 fuselage into the 747SP version. An extension to this process was carried out in the design of the A330 and 40 where a common wing was designed (Fig. 3.5).

The fuselage cross-section is of crucial importance in passenger comfort and cargo efficiency. Figure 3.8 shows a range of types including some of those proposed for the new 500–1000 seat ultra-high-capacity aircraft, typified by the A380.

One very important aspect of airliner design, which particularly affects take-off performance, is the question of noise. Earlier aircraft were judged to be too noisy and many aircraft were

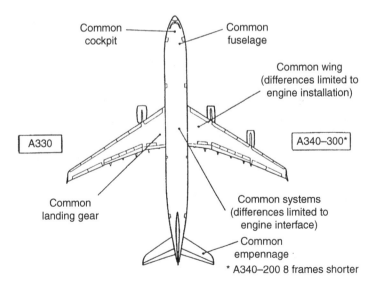

Fig. 3.5 Common features of the Airbus A330 and A340 aircraft.

Fig. 3.6 Boeing 787.

withdrawn from service because they did not meet new regulations. This was a costly exercise because many of these aircraft would otherwise have been in service for many years, although their older turbojet engines had poor fuel consumption. One compromise was to re-engine some aircraft with modern quiet turbofan engines and this was carried out for some DC8s and 707s. This process is being repeated by the use of new high-bypass turbofan engines for the Airbus A320 NEO and Boeing 737Max families. These changes offer significant fuel and operating cost benefits for programmes that have lower costs and shorter timescales, relative to totally new designs.

The supersonic airliner is faced with an even more severe problem in this respect. It was possible to reduce the take-off and landing noise levels of Concorde to those of contemporary jet aircraft, at its service-entry date, but these are unacceptable for any future supersonic aircraft. A further problem is that the noise produced by the sonic boom is inherent in supersonic flight. This is mitigated by restricting supersonic flight to over-sea portions of air routes. Future super-sonic transports (SSTs) are likely to follow this pattern and to use variable-cycle engines. Such engines will be relatively quiet turbo-fans on take-off, but have reduced by-pass ratio for increased

Fig. 3.7 Airbus A380.

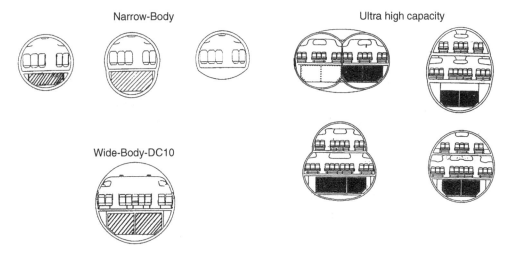

Fig. 3.8 Some airliner fuselage cross-sections (not to scale).

fuel efficiency at the likely cruise Mach number of 2.0 to 2.5. New project designs should triple Concorde's 100 seat payload, and provide trans-Pacific range.

Most of the foregoing discussion has concerned high-subsonic and supersonic jet transports, but regional aircraft are a crucial part of the aerospace business.

They have relatively short range and are sometimes called commuter or feeder-liners. This follows from their usual mode of operation, which is to carry passengers from large numbers of small airfields near small cities or towns to the larger cities, for transfer onto larger aircraft.

Fig. 3.9 EMBRAER 190.

They may be divided into three distinct categories. The largest carry 70 to 130 passengers at ranges of up to 1000 miles. They are usually turbo-fan powered and differ mainly from larger airliners only in having slightly reduced cruise speed, but improved field performance. The EMBRAER 190 (Fig. 3.9) and Bombardier C series (Fig. 3.10) are the main contenders in this market.

Mid-range regional aircraft are usually powered by turbo-props and carry 40 to 70 passengers, although Bombardier have recently modified their Challenger Executive jet into a 50-seat regional jet. The SAAB (Fig. 3.11) is a successful mid-range regional turbo-prop.

Smaller regional aircraft vary in capacity between 18 and 40 passengers, and most use turbo-prop engines.

The very smallest aircraft may operate from austere airfields and need to be very rugged. Requirements for performance do not require high speed, so externally braced wings and fixed landing gears may be used, as for the Cessna Caravan (Fig. 3.12).

3.2.2 Business or Executive Aircraft

There was a tremendous increase in airline activity in the late 1940s following the end of the Second World War. Business became much more international, but senior executives felt

Fig. 3.10 Bombardier C series.

Fig. 3.11 The SAAB 2000 regional turbo-prop with its stable-mate the SAAB 340.

themselves to be limited by the infrequent flights offered by airlines. Such flights operated from relatively few, large airports. This inflexibility led to the advent of smaller specialized aircraft, for use by senior executives. Early examples were the twin propeller-driven De Havilland Dove and

Fig. 3.12 Cessna Caravan regional turbo-prop, Berlin, 1996.

four-engined Heron aircraft. The former aircraft transported up to eight passengers over short flight sectors, but could use an extremely wide range of airfields because of its good airfield performance. Such aircraft were then augmented by jet-powered aircraft such as the De Havilland 125, which entered service in the mid 1960s. A developed version of this aircraft became the Raytheon-Hawker 800. Table 10.2 shows a decision-making chart which compares the performance of this aircraft against competing aircraft, in 1993. The aircraft shown are in what is termed the mid-range category. Maximum seating is eight, but typical occupancy is two to three. Cabin size is important for comfort in such small aircraft, and is improved by luxurious furnishings. Mid-range aircraft fly at airliner speeds, with ranges up to 2500 miles, but most flights are much shorter. Current costs are more than $10 million, so they are hardly toys for company executives and have to work to justify their existence. They do this by saving expensive executive time that would otherwise be lost in transit to, and waiting at, commercial airports. They can also act as in-flight meeting-rooms and as useful publicity enhancers by showing the company logo on their elegant exteriors. A number of longer-range aircraft have been developed, as well as conversions of smaller airliners.

The entry-level corporate jet has been developed with the aim of matching this price, but at improved comfort levels. Cranfield's E-92 project is shown in Fig. 3.13 and has the typical rear-fuselage-mounted engine configuration. The slightly forward-swept wing is *not* typical. It's aim is to give M 0.75 cruise speeds with a degree of natural laminar flow to reduce aerodynamic drag. The wing centre-section structure is not a cabin obstruction and the carbon fibre wing structure saves weight and prevents aeroelastic problems.

Fig. 3.13 CAD model of Cranfield E-92 executive jet project, 1993.

3.2.3 Cargo Aircraft

These have always been the 'Cinderella' of aircraft types. Air cargo is expensive compared with sea freight and attempts are made to cut costs. A very large proportion of air cargo is carried in airliners along with passengers. This cuts costs and gives a good service frequency. The wide-body aircraft are particularly useful in that they have large under-floor holds that can be filled with standard containers (Fig. 3.14). Larger items of cargo, however, must be carried above the floor and so large doors are cut into the sides of the fuselage.

For even larger payload dimensions, more drastic conversions are required. The most capable civil freighter is the Boeing 747F, (Fig. 3.15) which uses a large visor nose door, together with a fuselage side door. The use of standard containers in such large numbers drastically reduces freight costs. Figure 3.16 shows the Boeing 747-8 latest version of the aircraft, which is capable of carrying some 150 tonnes of cargo.

The pure cargo aircraft has somewhat different requirements from the airliner since the payload must be large over a long range, but speed is not of primary importance. Since the volume capacity is often as great a consideration as weight capacity, the fuselage of the cargo aircraft tends to be large and blunt, with very large doors. Low speed performance is sometimes of considerable importance so as to give the aircraft the ability to operate from a large number of places where only second-rate facilities are available. This not only means large area wings with refined high lift

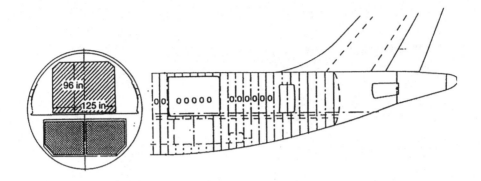

Fig. 3.14 Airbus A300 main deck cargo conversion, 1995.

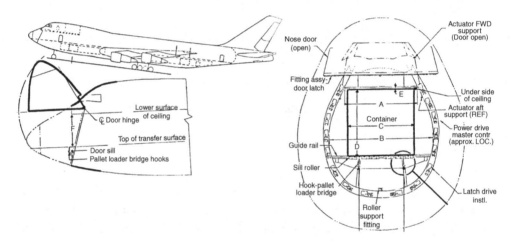

Fig. 3.15 Boeing 747-200F cargo conversion, 1980.

devices, but also large and heavy undercarriages. Two older aircraft, the Viscount 810 and Argosy had almost identical power plants, but the latter could carry up to twice the payload with a 20% reduction in cruising speed and has a 50% greater wing area (Fig. 3.17).

Military cargo aircraft are more specialized and are described later.

3.2.4 Agricultural Aircraft

Small fixed-wing aeroplanes or helicopters perform extremely useful duties to support farming operations throughout the world. They are particularly useful for pest control or fertilizer dispersal in areas of the world where surface access is difficult. They are therefore used in many developing countries. Their flexibility and efficiency is thus an important element in the world's agricultural

Fig. 3.16 Boeing 747-8 Freighter.

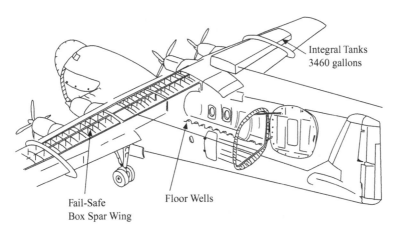

Integral Tanks
3460 gallons

Fail-Safe
Box Spar Wing

Floor Wells

Fig. 3.17 Armstrong-Whitworth Cargo Aircraft, 1966.

economy. Most aircraft are rugged light aircraft typified by the PZL-WILGA shown in Fig. 3.18. This has a piston engine and a chemical hopper near its CG. This is to minimize pitch changes when up to 50% of the aircraft mass is rapidly discharged. The WILGA and other agricultural aircraft have good take-off and low-speed performances, to give safety close to the ground. Other more recent agricultural aircraft have moved to the low-wing configuration, with more payload and

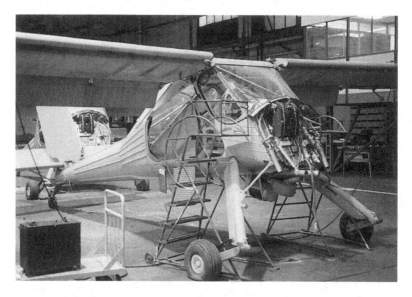

Fig. 3.18 Manufacture of PZL–WILGA agricultural aircraft, 1994.

Fig. 3.19 The Cranfield Al aerobatic aircraft, 1978.

the pilot seated behind the wing. This is to prevent him or her being sandwiched between the heavy hopper and the engine, if the aircraft were to fly into the ground. Moving the pilot aft means that the cockpit needs to be raised to aid pilot vision. The pilot is protected by a rollover crash cage and the whole region designed to resist 25 g axial decelerations. Those requirements lead to the

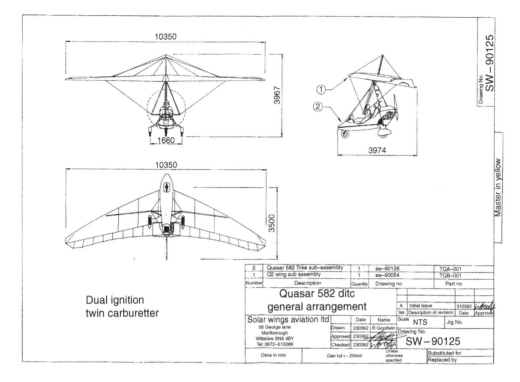

Fig. 3.20 The Solar Wings Quasar 1995 (dimensions in mm).

characteristic hump-back shape of many agricultural aircraft. The most widely used aircraft have a payload of half a tonne and are powered by single 200–300 hp piston engines. Larger aircraft may carry twice the payload and be powered by ageing 600–700 hp radial engines or by more expensive turbo props. The agricultural aircraft sometimes provide the last significant role for the biplane. Such aircraft are very strong and have very good low-speed manoeuvrability.

3.2.5 Light Aircraft

This is the term used to describe the smallest size and most numerous aircraft. Many thousands of aircraft such as the Cessna 150, Beagle Pup, etc. are used for pleasure, pilot training and very light transport. They are of simple and cheap construction, usually powered by a single piston engine. One specialized category is that of aerobatic aircraft such as the Cranfield Al, which has to have a very strong construction to withstand severe manoeuvring loads (Fig. 3.19). Micro-light aircraft developed from powered hang-gliders and have their own airworthiness requirements, as do gliders.

The simplest micro-lights use the pilot's weight-shift for flying control, such as the Solar Wings Quasar (Fig. 3.20).

CHAPTER 4

WHY IS IT THAT SHAPE?:
OTHER TYPES

4.1 MILITARY AIRCRAFT TYPES

4.1.1 Fighters

These aircraft have had several designations over the years, including pursuit and interceptor aircraft. There are many types, and definitions become blurred between fighters and bombers and ground-attack aircraft. True fighters may be divided into two major classes. The first class is that of relatively simple short-range interceptor aircraft (class 1), whilst the second used to be called 'all-weather' fighters which tend to have longer range and more avionic equipment (class 2). Figure 4.1 shows a plot of combat wing loading against thrust/weight ratio, which is a good indication of the manoeuvrability of combat aircraft (see Chapter 3 for the definition of thrust/weight and wing loading). Recent requirements for high agility have moved aircraft towards the top left-hand corner where high thrust/weight and low wing loading improve climb, sustained turn and attained turn performance. Most aircraft in this region are class 1 short-range fighters and class 2 are closer to the bottom-right quadrant, where the higher wing loading leads to better long-range cruise.

The *class 1* short-range, high-performance interceptor carries the minimum of equipment and maximum speed is always important. However, for such an aircraft the rate of climb and manoeuvrability may be even more important. This arises from the fact that due to the fighter's necessarily short range it cannot take off until a target is definitely located, but must then rapidly climb to interception. A very high thrust/weight ratio is implied by these requirements, particularly at altitude. Compatible with reasonable operational safety, everything must be placed second to climb and interception performance. Only one crew member can be carried and acquisition and life-cycle costs are reasonably low. Due to the relative simplicity of the equipment carried, reliability should be high. The Typhoon is in this category (see Fig. 4.2).

The high thrust/weight ratio for combat manoeuvrability leads to the thought that only a little extra thrust might give vertical take-off and landing capability (VTOL). The BAE SYSTEMS Harrier has been developed into an extremely effective strike/bomber aircraft with the very significant operational advantage of lack of reliance on runways. The invention of the ski-jump runway improved payload/range by offering an extremely short take-off distance (STO). These innovations led to the extremely potent Sea Harrier interceptor aircraft flown from small aircraft carriers.

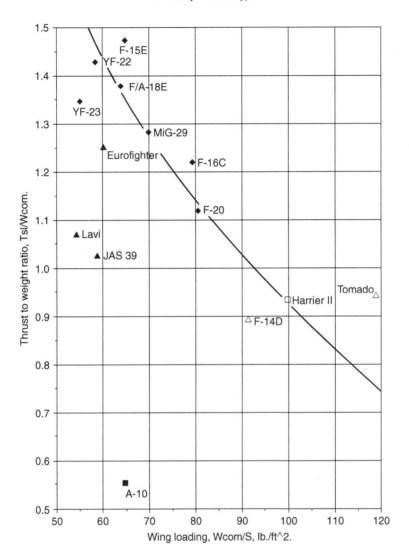

Fig. 4.1 Fighter aircraft wing loading and thrust trends.

The US Navy proposed a supersonic V/STOL interceptor requirement and this specification was examined by the College of Aeronautics to produce the S-83 design (Fig. 4.3).

A close-coupled canard arrangement was chosen because it gave improved lift characteristics for short take-off and combat. The foreplane made possible the achievement of high angles of attack, and improved area-ruling for reduced supersonic drag. Negative subsonic longitudinal stability combined with active controls gave improved aerodynamic performance.

The forward-swept wing has an aspect ratio of 4 and utilizes a 5.9% thick super-critical aerofoil section. Claimed advantages of such a configuration are:

Fig. 4.2 Eurofighter Typhoon, 2013.

  (i) Configuration flexibility.

  (ii) Significant higher manoeuvre lift/drag ratio.

 (iii) Lower trim drag.

 (iv) Lower stall speeds and slower landings.

  (v) Virtually spin proof.

 (vi) Better low-speed handling.

The aircraft uses the remote augmented lift system (RALS) developed by the General Electric Company. Two propulsion units are used mounted side by side in the rear fuselage, having variable cycle capability. A double bypass split fan provides airflow to the single remote augmentor nozzle during vertical take-off and landing. Primary exhaust is through augmentor deflector exhaust nozzles (ADEN).

The remote augmentor nozzle provides thrust under the forward fuselage. Thus the aircraft rises on three points of thrust. Canards have been successfully used on non-V/STOL aircraft such as Rafale, Viggen, Gripen and Eurofighter. The late 1980s/early 1990s saw the conceptual designs of several advanced V/STOL aircraft, an example of which is the Cranfield S–95.

The *class 2* 'all weather fighters' are designed to have longer-range patrol ability. Such aircraft tend to be large, heavy, complex and expensive. The equipment carried has to be extensive in order to ensure satisfactory interception in bad weather and a crew of two may be required to

Fig. 4.3 Model of Cranfield S-83, 1994.

operate the aircraft. Rate of climb is of lesser importance since the aircraft should be patrolling at altitude when alerted, but thrust/weight ratio will still be high for manoeuvres. Fuel economy for maximum duration of patrol is critical, which is one of the reasons for the use of low-bypass turbo-fan engines. These are usually fitted with afterburners for good combat thrust/weight ratio. Attempts to design one aircraft to fill both short- and long-range roles have usually resulted in mediocre aircraft not suited properly to either task. In choosing between the two it is necessary to consider economics, geography and any other defensive means which may be available, and the actual nature of the enemy threat. Stealth technology can be used to improve the aircraft surviv-ability, as in the F-22 shown in Fig. 4.4.

Non-V/STOL *naval* fighters are similar in concept to land-based aircraft, but additional restrictions and requirements are imposed on their design. Apart from the obvious sacrifice of performance and safety necessary to enable operation from a restricted deck area, the limited storage space in an aircraft-carrier implies a considerable restriction of the total number of aircraft which can be carried. Therefore each type should be able to fulfil more than one task and reliability is of prime importance. In addition to its defensive capacity, the longer range type of interceptor becomes an offensive aircraft, so that it can be used in a strike role.

The limited storage space and deck elevator dimensions usually require wing and perhaps fuselage section folding. Additional complications arise from strengthening for catapult launch and arrestor-wire hook attachments. The deck of the aircraft carrier is a moving runway that will pitch

Fig. 4.4 F-22 fighter (with author), 2000.

and roll. Combined with arrested landings, this is very punishing for landing gears, which need to be greatly strengthened. The salt-laden atmosphere restricts the use of certain structural materials that may be used on land-based aircraft.

All of these complications lead to weight and maintenance cost increases, which make naval aircraft design one of the most challenging aircraft design tasks.

4.1.2 Bombers

Most fighter aircraft have bombing capability, but many air forces have a need for dedicated bomber aircraft. These aircraft can vary enormously in size according to their particular tactical or strategic roles. Examples of extremes in this respect are the Douglas A4D which weighs only 8 t and the Boeing B52 with a take-off mass of more than 220 t. However, each type is designed with a common over-riding consideration – the ability to deliver a warhead at a given range with a minimum chance of loss due to enemy action.

Although the range requirement and warhead size tend to decide the size of the aircraft, the method adopted to reduce the chance of enemy interception also has a big repercussion on the design. If high speed is chosen for this purpose, as in the Rockwell Bl which flies supersonically, the range will be limited because of the high fuel consumption, or the size of the aircraft may be prohibitive. The same is true if low-level attack is adopted so that the approach to the target can be made under the enemy's radar screen. On the other hand, subsonic flight at very high altitude requires a large wing area and hence a heavy structure weight. A further possibility is to defend the bomber with guns, as in many former Soviet aircraft, but since the advent of the guided missile this technique has fallen into disfavour. It has been replaced by the concept of a 'stand-off' bomb or cruise missile which is launched from outside the main target area. It is impossible to draw any general conclusions about the final configuration in view of the number of possibilities, except to say that the bomber is a complex, and usually large, aircraft. Cruise performance is critical of course, but, like the airliner, low speed performance may well force the designer into a compromise, or the use of variable geometry, as in the strike version of the Tornado.

4.1.2.1 *Reduction of the Probability of Being Detected*

High-flying aircraft have become vulnerable to high-performance surface-to-air missiles (SAMs) and interceptors, so new defensive techniques have been developed:

(i) Electronic measures – A modern bomber defends itself with electronics rather than with guns or missiles. A radio frequency surveillance/electronic countermeasures (RFS/ECM) system detects hostile radio signals and automatically protects the aircraft against them, making it difficult to locate and hit with missiles. ECM operates in one of two ways. The first is to send a high-powered 'blanket' of signals on many frequencies. The second sends decoy signals to mislead the opposition.

(ii) Stealth aircraft – This is a relatively recent concept in which the aircraft is designed to be almost invisible to radar and infra-red (IR) sensors. Radar cross-section (RCS) can be reduced by careful design of the airframe to reduce sharp corners and hide the engine intakes from the ground. Special materials have been designed to absorb radar signals and they are placed in critical areas of the aircraft.

Figure 4.5 gives a comparison of the radar cross-sections of a number of bomber and fighter aircraft [8].

Figure 4.6 shows the configuration of a supersonic bomber aircraft designed to minimize radar cross-section [9]. The high altitude cruise requirement was M 1.6, with M 0.9 at low altitude. Its main features were:

(i) Wing – A highly swept, basically delta wing with rounded tips was chosen. This combines good supersonic wave drag and favourable stealth behaviour. The wing is mounted as low on the body as allowed by other considerations to minimize the wing-body blending needed to avoid introducing corner reflectors. Stealth considerations dictated that leading edge devices should not be employed, but the inner leading edge is swept forward to form a chine along the sides of the forward fuselage. This has the consequence of eliminating vertical fuselage surfaces and providing a shielding to the cockpit canopy and engine air intakes, as well as enhancing wing-body lift.

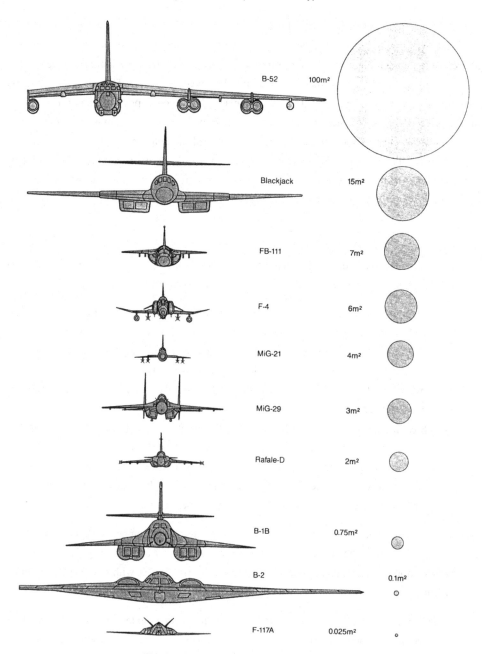

Fig. 4.5 Relative radar cross-section, 1990.

(ii) Body – All weapons are carried in a bay which is located beneath the wing structure. The somewhat large canopy was chosen to give good visibility to the crew. This led to the need to shield it as much as possible, although it was proposed to give the cockpit canopy a non-reflective metallic coating.

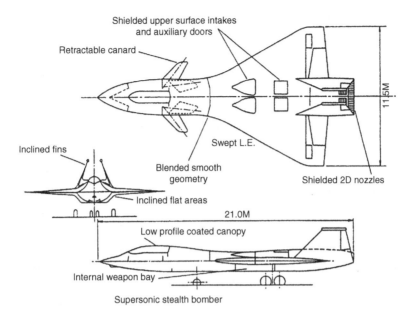

Fig. 4.6 Supersonic stealth bomber configuration, 1987.

(iii) Powerplant installation – The two powerplants are buried in the fuselage behind the wing carry-through structure to minimize frontal RCS. The air intakes are located on the upper surface of the blended wing-body and are thus largely shielded from the front and below. Although similarly shaped in plan to flush intakes they do in fact face forward and slightly upwards. Some reliance is placed on the leading edge vortex to feed air into the intake region. Auxiliary doors are provided for low speed, the main intakes being of fixed geometry optimized at M 0.9. The intake ducts are long, curved and lined with radar absorbent material (RAM). Two-dimensional exhaust nozzles are located on the upper side of the extreme end of the fuselage.

(iv) Fins (vertical stabilizers) – The vertical stabilizing area is split into two to minimize side area and the pair of surfaces canted inwards from the aft end of the wide fuselage. In this configuration the fins provide a side shield for the exhausts and the adverse effect of corner reflection at the wing junction is minimized.

(v) Canards – Low-speed take-off and landing performance dictated the need for a small canard lifting surface. Although this could be used to provide gust alleviation during low-level flight it is not required for performance or control at high subsonic and supersonic speeds. The canard is a significant penalty for stealth considerations, so it is proposed that it be retracted into the fuselage sides for normal cruising flight.

(vi) Overall considerations – Much of the construction employs composite materials. Care has been taken to avoid cutouts and discontinuities on all upper surfaces. Lower access doors would have to be sealed and bonded. The main landing gear retracts into the wing-body blend fillet alongside the weapon bay and does not present a frontal discontinuity. Nosewheel retraction is conventional. Powerplant access is difficult and the high engine thrust line introduces a trim drag penalty, but these matters are part of the price that must be paid for low observability.

Infra-red sensors are used extensively on missiles that home onto hot pipes and exhaust plumes. These sources can be reduced by the use of turbo-fan engines and cooling or shielding of the exhausts.

4.1.2.2 Bomber Fuselage Layout

Most dedicated bomber aircraft have an internal weapon bay and its location dominates each bomber's configuration. The prime consideration is the location of the weapon load, which must be disposed on or about the CG. Thus, unless an unusual wing arrangement, such as a canard, is used, the wing structure intersects the fuselage in the region of the weapon bay. The wing must therefore have a mid or high location. A mid wing may be used in some cases, but the need to make maximum use of fuselage volume for equipment or fuel often makes this difficult, and in this case a high wing is preferred, as this can be built across the top of the fuselage. Figure 4.7 shows the S-3 anti-submarine aircraft, which illustrates this arrangement.

The crew must normally be in a pressurized compartment and they are therefore grouped close together near the nose of the aircraft. Many of the smaller strike bombers carry only two crew, who may be arranged in either a side-by-side or a tandem cockpit. There is a difference of opinion about the relative merits of these arrangements. Side-by-side should give enhanced

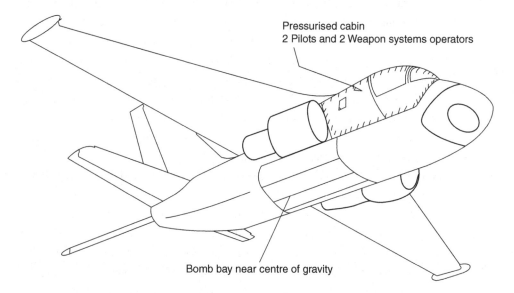

Pressurised cabin
2 Pilots and 2 Weapon systems operators

Bomb bay near centre of gravity

Fig. 4.7 S-3 Viking anti-submarine aircraft.

operational efficiency and an easier design if an escape capsule is used. On the other hand, a tandem cockpit is usually much easier to fit into the overall fuselage layout, especially if the weapon bay is long and relatively narrow. It also enables more equipment to be stored within direct reach of the crew. This is desirable both from an operational and a weight point of view. It is generally preferable to keep as much equipment as possible in the fuselage adjacent to the cockpit.

The layout in the region of the bomb bay must be carefully considered to ensure as far as possible that no undue turbulence will occur when the bomb doors are open and that the bombs can be released safely. The rear of the bomb bay must be designed to collapse readily under water impact loads in the event of ditching, as otherwise the aircraft will nose over. External carriage of stores is worthy of consideration, since the associated drag may be offset by the greatly reduced drag after release of the stores and the elimination of the complexity associated with a bomb bay.

4.1.3 Ground Attack/Trainer Aircraft

Aircraft play a major role in any land war. Central to their role are laser-guided weapons, which are especially suited to aerial use because aircraft have an overall view of the battle area and can easily carry the required laser. The laser emits a narrow light beam, with precise characteristics that may be hard for an enemy to interfere with. Once the target has been spotted, it can be designated by aiming a laser at it. The laser may be in the aircraft carrying the weapons, in an accompanying aircraft (or small remotely piloted vehicle), or even aimed by a soldier on the ground. Today's missiles can be made to fly towards the light diffused or scattered from the target, and do so with high reliability and great accuracy. What this means is that, in conjunction with modern sighting systems, aircraft can so dominate a land battle that it becomes virtually an air battle, as shown in the 'Desert Storm' operations.

Targets for such weapons are tanks and armoured fighting vehicles, which may also be attacked at closer range by modern cannon, rockets or cluster bombs.

There remains, as always, the problem of aircraft vulnerability. Until the 1970s, most tactical aircraft could usually be brought down, even by small arms if strikes were taken in a vital area. The pilot has been armoured since the 1930s, and fuel tanks may be protected by self-sealing or purged with nitrogen or other inert gas. However, the basic aircraft has been so penetrable by bullets and so complex that even simply equipped troops could be dangerous to aircraft flying really low, while fast-firing cannon and small surface-to-air missiles were potentially lethal at ranges of a mile or more. Much thought therefore went into making tactical aircraft 'survivable'.

Two of the important ground-attack aircraft are the BAE SYSTEMS Harrier (Fig. 4.8) and the American Fairchild A-10A. The former only needs a flat, level patch of ground as an airfield and is thus often able to reach frontline troops within seconds of a call for help. It carries all normal weapons and a laser designator, and has the advantage of presenting an extremely small, agile, and elusive target to an enemy. The A-10A is much bigger, and is a conventional straight-wing

Fig. 4.8 BAE SYSTEMS Harrier.

A-10 Survivability features

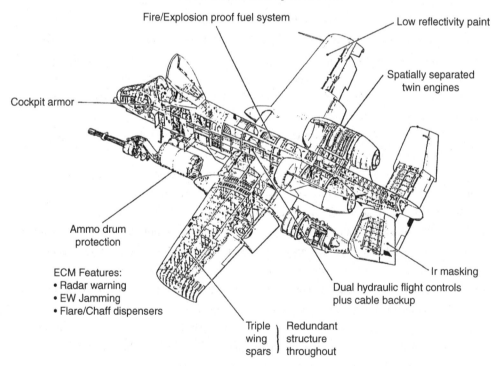

Fire/Explosion proof fuel system

Low reflectivity paint

Spatially separated twin engines

Cockpit armor

Ammo drum protection

ECM Features:
• Radar warning
• EW Jamming
• Flare/Chaff dispensers

Ir masking

Dual hydraulic flight controls plus cable backup

Triple wing spars Redundant structure throughout

Fig. 4.9 Fairchild A-10.

aeroplane with a take-off run of 900 m (3000 ft), or rather more at full load. It was designed around a large 30 mm cannon, bigger than any other currently in use, and it can also carry about 8 t of weapons of many kinds. Figure 4.9 shows this aircraft and emphasizes its survivability features.

Less capable ground attack aircraft are almost invariably developed from the current military training aircraft, which are discussed below. Such aircraft need to have weapon training capability and are useful dual-purpose aircraft for smaller air forces.

Military jet trainers differ from their predecessors in that if the seats are in tandem the instructor is raised at least a foot above the pupil, giving him a better view, especially on the landing approach when the aircraft will fly in a nose-up attitude. Both occupants sit in ejection seats, and most trainers may be equipped with at least a simple sight, a fixed gun and racks for light bombs or rockets.

Nearly all single-seat attack fighter aircraft are also made in two-seat forms for use as conversion trainers, which provide supersonic and operational training.

The BAE SYSTEMS Hawk trainer has been developed into a capable single-seat light fighter, whilst the single-seat Italian/Brazilian AMX ground attack aircraft has been modified in the reverse direction to provide two seats. Larger aircraft, such as the BAE SYSTEMS Jetstream, are used for multi-engine pilot training and navigator training.

4.1.4 Reconnaissance Aircraft

The earliest reconnaissance aircraft were used during World War I for the direction of artillery and for photography. The importance of such aircraft has continuously increased and they operate in many different areas.

4.1.4.1 Electronic and Photographic Missions

In the 20 years following World War II, electronic reconnaissance was of enormous importance. The scale of operations can be guessed from the number of aircraft, mainly of the US Air Force and Navy, that were shot down. Aircraft types included the RB-57 and the Lockheed U-2. The second of these, a specially designed ultra-high-altitude aircraft, made flights over Soviet territory for four years until 1960, when a U-2 was brought down by a Russian surface-to-air missile.

Today's reconnaissance missions are flown by a wide variety of aircraft and satellites. For strategic missions at the highest possible altitude and speed the US Air Force Lockhead S-71A Blackbird was used. The nearest Russian equivalent is the reconnaissance version of the MiG-25 Foxbat, although this has a much shorter range. Military satellites are widely used for reconnaissance.

4.1.4.2 Tactical Missions

Tactical reconnaissance is flown either by tactical attack or fighter aircraft carrying multi-sensor pods on external weapon pylons, or by special versions with the equipment housed internally, but the latter have few weapons. This type of mission may be extremely risky, but advances in electronics enable the use of remotely piloted vehicles. These aircraft are very small, cheaper than manned aircraft, and do not risk a pilot.

4.1.4.3 Maritime Missions

One of the major strategic weapon systems is the nuclear-powered submarine armed with ballistic missiles. Very sophisticated aircraft such as the BAE SYSTEMS Nimrod are required to counter this threat. This is a much converted version of the Comet airliner, with a larger crew and a vast array of sensors and weapons. Sensors include radar, magnetic anomaly detectors (MAD), sono-buoys and equipment that will 'sniff' fumes of diesel-powered submarines. This aircraft combines a high-speed dash capability to reach a target quickly and long endurance on station.

Advances in avionics have meant that conversions of much smaller commercial aircraft can make very effective maritime reconnaissance aircraft.

4.1.3.4 Airborne Early Warning (AEW) Missions

Ground-based radar systems always suffer from the effects of the curvature of the Earth. This leaves a 'blind spot' that is exploited by very low-flying aircraft. The American version of this aircraft is the AWACS derivative of the Boeing 707, which mounts a large rotating antenna on top of the fuselage.

Other aircraft have been used to perform this role, most with the rotating dish antenna. One exception was the unsuccessful AEW version of the Nimrod. This utilized fore and aft scanners in nose and tail to give 360° coverage (see Chapter 11).

4.1.5 Military Cargo Aircraft

The fuselage cross-section is critical on all dedicated cargo aircraft, but it is particularly important in military aircraft when large tanks are to be carried. Figure 4.10 shows a number of military cargo aircraft cross-sections.

Figure 4.11 shows a side-view of the Antonov-124, which embodies most of the desirable features of dedicated cargo aircraft:

  (i) Unobstructed cargo hold.
  (ii) Full-width doors, with hinged ramps to give approx 11° drive-on angle.
  (iii) Level floor, suitably reinforced.
  (iv) Adjustable undercarriage to give 'truck' bed height of rear ramp.

This aircraft uses a high wing which is usually chosen for freighters because it leaves the fuselage floor close to the ground to ease loading. This leads to pod-mounted undercarriage legs, which can cause aerodynamic problems.

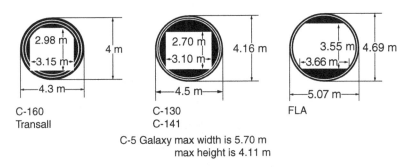

C-160
Transall

C-130
C-141

FLA

C-5 Galaxy max width is 5.70 m
max height is 4.11 m

Fig. 4.10 Military cargo aircraft cross-sections.

Fig. 4.11 Antonov-124 cargo aircraft, 1996.

In early 1972 the USAF requested proposals for an advanced medium STOL (short take-off and landing) transport (AMST), primarily as a replacement for the C-130. It was also planned to gather all the diverse new technology of increasing wing lift at low airspeeds and the possibility of integrating wings and propulsion systems to achieve better short-field performance. The two finalists were the McDonnell Douglas YC-15, and Boeing YC-14. Boeing chose to use a challenging, but promising method of powered lift called upper-surface blowing (USB). The two large turbo-fans exhaust through a flattened nozzle above the upper surface of the wing. In the high-lift regime the inboard flaps are deflected to 85° with sliding link-plates forming a smooth and continuous upper surface, while vortex generators emerge above the wing. The engine efflux passes over the wing and curves down over the flaps, giving lift almost equal to the net engine

Fig. 4.12 Boeing C-17, 1996.

thrust. The outboard flaps are large double-slotted units, and the outboard leading edge is a variable-camber slat blown by air bled from the engines.

McDonnell Douglas chose externally blown flaps (EBF), in conjunction with well-proven transport engines. The four JT8D engines exhausted through multi-lobe mixer nozzles directly under the wing lower surface so that when the powerful titanium double-slotted flaps are lowered, the whole engine efflux is diverted downwards. With a much increased flow of entrained air, this gives sufficient lift for STOL and for flight at speeds below 100 mph(160 km/h).

Many of the features of the YC-15 were incorporated on the later, larger C-17 cargo aircraft shown in Fig. 4.12.

4.2 ROTORCRAFT, V/STOL AIRCRAFT AND UNINHABITED AERIAL SYSTEMS

The main difference between a rotorcraft and a fixed-wing aircraft is the principal source of lift. The fixed-wing aircraft derives its lift from a fixed aerofoil surface, while the rotorcraft derives lift from a rotating aerofoil called the rotor. Aircraft are classified as either fixed-wing or rotating wing; most of the latter are helicopters, but auto-gyros and tilt-rotors are important members of the rotorcraft family. Auto-gyros use an unpowered rotor for lift, but use conventional propulsion for forward flight. An auto-gyro cannot hover, as it requires forward speed to spin the rotor. Tilt-rotors will be discussed later.

During any kind of horizontal or vertical flight of a helicopter, there are four main forces acting on the aircraft: lift, thrust, drag and weight. These forces are identical to those applied to fixed-wing aircraft, but the lift and thrust act in different ways. These are shown in Fig. 4.13.

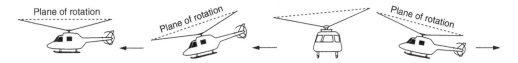

Fig. 4.13 Helicopter motion.

During hovering flight in a no-wind condition, the tip-path plane is horizontal. The lift and thrust forces both act vertically upward. Weight and drag both act vertically downward. When lift and thrust equal weight and drag, the helicopter hovers.

For forward flight, the tip-path plane is tilted forward thus tilting the total lift-thrust force forward from the vertical. In sideward flight, the tip-path plane is tilted sideward in the desired flight direction thus tilting the total lift–thrust vector sideways.

For rearward flight, the tip-path plane is tilted aft, tilting the lift–thrust vector rearward. The thrust component is rearward and drag forward, just the opposite to forward flight. The lift component is straight up and weight straight down.

In non-hovering flight, the rotational speed of the rotor is added to, or reduced by, the aircraft's speed. This leads to unbalanced lift and forces, which are compensated for by careful rotor hub design. As the main rotor of a helicopter turns in one direction, the fuselage tends to rotate in the opposite direction, due to the torque produced from the rotor. The force that compensates for this torque and provides for directional control can be produced by means of an auxiliary rotor located on the end of the tail boom. This auxiliary rotor, generally referred to as a tail rotor, or anti-torque rotor, produces thrust in the direction opposite to torque reaction developed by the main rotor. Alternative anti-torque remedies include contra-rotating co-axial rotors or twin counter-rotating rotors on wing tips, or nose and tail, as in the Boeing Vertol Chinook.

A recent research development has been the NOTAR (NO Tail Rotor), which produces high-energy reaction air at the tail, ducted from the engine. This resists the main rotor torque and dispenses with a tail rotor.

Helicopters are inherently more complex and more expensive both to buy and to operate, than fixed-wing aircraft with the same payload. They are also limited in speed and range, but their great versatility means that they can do jobs impossible for fixed-wing aircraft. The ability to hover and VTOL performance are particularly strong features.

There has been rapid growth in the construction and operation of helicopters by armies, navies, air forces and civil operators in many roles described below.

4.2.1 Civil Helicopter Operations

Experimental mail operations started just after World War II and led to commercial helicopter operations. There have been operations linking city centres with major airports, and many successful routes, such as that from Cornwall to the Scilly Isles in Great Britain. Airline operators require the use of relatively large twin-engined aircraft such as the Sikorsky S-61, Chinook and Puma. Helicopters are also used for specialized freight operations, but bulky loads have to be slung externally.

A large helicopter growth area in the UK is the carriage of passengers and freight to North Sea oil platforms. Modern helicopters are used for executive transport, police work, forestry, etc. One major use is in agricultural work. Crop spraying can be performed in places inaccessible to other aircraft. The downwash from the rotor has the advantage of helping insecticide penetration by exposing the underside of leaves. As in other operations, helicopter cost per acre sprayed is considerably higher than for fixed-wing aircraft.

4.2.2 Military Helicopter Operations

Military helicopters were first used in significant numbers during the Korean war. Their flexibility and ability to dispense with runways overcome their limitations in range, speed and payload-carrying capability, in a number of important operational scenarios. The range of roles has increased significantly and reached its peak during the Vietnam war. Most modern armies utilize large fleets of helicopters for many purposes. The initial role of casualty evacuation was augmented by using helicopters to carry squads of infantry soldiers rapidly to remote areas. In this role they are termed assault helicopters. One of the most capable is the Aerospatiale Puma, from which Denel Aviation, South Africa, developed the Oryx Medium Transport Helicopter. Denel then developed and manufactured the Rooivalk attack helicopter, which has some commonality of parts with both the Puma and the Oryx (Fig. 4.14). This approach has benefits in development and maintenance costs for customers operating both types of helicopter. The Rooivalk is a typical modern attack helicopter, which utilizes a slim, heavy-armoured fuselage. The gunner sits in the front cockpit to aim the nose-mounted cannon, and rockets and missiles attached to the stub-wing. The pilot sits in the raised rear cockpit, where he has good vision. The helicopter is fitted with twin turbine-engines, which drive the main and tail rotors, by means of gearboxes and drive shafts. Such aircraft operate close to the ground and can be used in anti-tank or anti-personnel roles. They are susceptible to ground fire and are thus designed to be as invulnerable as possible. The engines have heat-shields to deter infra-red missiles and the fuel system uses many protective measures. Bigger helicopters are used to transport larger number of troops and equipment. It is possible to carry externally slung loads, such as artillery pieces.

Many navies of the world use helicopters in a number of roles. Warships as small as frigates have a helicopter landing-pad near their sterns and there are a number of helicopter aircraft carriers. Conventional aircraft carriers have a mixed complement of fixed-wing aircraft and helicopters. The anti-submarine warfare (ASW) role is one of the most vital. A small crew operate a wide range of sensors and carry depth charges, torpedoes, rockets or missiles. They can thus be used to detect and destroy submarines or small surface vessels. They also perform vital rescue and

Fig. 4.14 Denel Rooivalk attack helicopter, 2016.

transport tasks. Westland have developed an airborne early warning (AEW) version of the Sea King helicopter to rectify the lack of such protection during the Falklands war. The Westland AGUSTA EH-101 Merlin helicopter has civil, army and naval variants, and has the advantage of using three turbo-shaft engines to drive the rotors.

4.2.3 Other VTOL Aircraft

No discussion of rotorcraft is complete without reference to VTOL and STOL aircraft, i.e. aircraft designed for vertical or short take-off and landing, respectively. No matter how the additional lift is obtained, it implies extra power relative to a comparable conventional aircraft. Unfortunately this means increased noise and cost in aircraft of economic size. Thus, although the helicopter has a proven use for such duties as rescue and short range communications, the application of the VTOL aircraft to wider civil operation appears to be faced with considerable obstacles. STOL undoubtedly has applications in the small transport aircraft. The DHC-7 four turbo-prop-engined fixed-wing airliner has had considerable success.

Many prototype V/STOL aircraft have flown, but very few have been totally successful. The Harrier fighter, which has already been mentioned, is an obvious exception. A promising early design was the Fairey Rotordyne and an aircraft of more recent interest is the tilt-rotor V-22 Osprey.

Tilt-rotor aircraft combine the flexibility advantages of the helicopter and the speed and range advantages of fixed-wing aircraft. It remains to be seen if these advantages can overcome the

Fig. 4.15 Cranfield tilt-rotor project.

Fig. 4.16 Micro-light autogyro.

inevitable complexity and cost disadvantages of the tilt-rotor aircraft. Jonkers [10] suggested a design of an executive aircraft. A later concept is shown in Fig. 4.15. This has engine nacelles that tilt to give vertical lift, or forward-flight thrust. The possibility of single engine failure on take-off has led to the adoption of engine-to-engine cross-shafting, so that the remaining engine may drive both rotors. This eliminates potentially catastrophic asymmetric roll and yaw motion.

The auto-gyro is another important rotorcraft category. The rotor is unpowered, but once it turns it produces lift and becomes a rotary-wing. Propulsion is provided by a conventional engine. It is simpler in construction than a helicopter, but it cannot hover.

Figure 4.16 shows a micro-light auto-gyro.

4.2.4 Uninhabited Aerial Systems (UAS)

There has been significant growth in this sector over the past 20 years, with the importance of such systems predicted to grow rapidly in the future. Many companies, universities and other organizations have produced UAV aircraft, as many are relatively simple and cheap to develop. A simple rationale for UAVs is that they are suitable for "dull, dirty or dangerous" missions. Dull missions include reconnaissance and data relay UASs, which operate in both civil and military markets; "dirty" missions include nuclear reconnaissance, whilst uninhabited combat air vehicles (UCAVs), are clearly dangerous missions. The following list of UAS categories is not exhaustive, but it reflects the major types, and illustrations are given of projects with which the author has been associated.

4.2.4.1 Civil/ Military Reconnaissance systems
Figure 4.17 shows the Cranfield Observer technology demonstrator UAV, typical of short-range tactical reconnaissance aircraft. It uses multiple video cameras in the nose, is launched by catapult and recovered by parachute and airbag. Figure 4.18 shows a medium-altitude, long-endurance (MALE) UAV, which cruises at some 10 000ft, for 10–12 hours.

Figure 4.19 shows the Cranfield U-2000 project which is a much larger HALE aircraft (high-altitude, long-endurance). It could fly at altitudes of some 65 000 ft, for more than 42 hours.

4.2.4.2 Uninhabited Combat Air Vehicles (UCAV)
UASs can be used for reconnaissance, and can be used as target aircraft or decoys, but the growing application area is that of armed UCAVs. They include ground attack aircraft, such as the General Atomics Reaper (Fig. 4.20)

Fig. 4.17 Cranfield Aerospace Ltd. Observer.

Fig. 4.18 BAE SYSTEMS Mantis.

Fig. 4.19 Cranfield U-2000 CAD image.

Fig. 4.20 General Atomics Reaper.

Fig. 4.21 Cranfield U3 CAD image.

UCAVs may be developed into fighters in the future, but current efforts are going into developing UCAV bombers, such as the Cranfield U-3 (Fig. 4.21)

4.2.4.3 Flying Demonstrator UASs

UASs can be a relatively quick and cheap way of demonstrating new and potentially risky technologies in a realistic environment. This can lead to development risk reduction and can greatly enhance the development of production aircraft.

Fig. 4.22 Boeing/NASA/Cranfield X-48C.

Fig. 4.23 Cranfield Kestrel.

There has been considerable interest in the possibility of developing blended-wing-body aircraft, and Cranfield University has produced three flying demonstrator UAS aircraft. Two Boeing/NASA X-48B and C demonstrator aircraft were designed and built by Cranfield Aerospace Ltd., using Boeing data (Fig. 4.22).

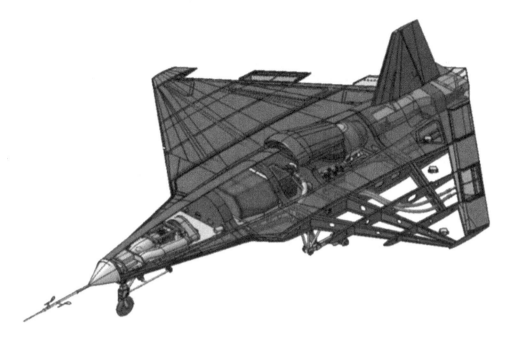

Fig. 4.24 Cranfield Demon CAD image.

Fig. 4.25 Cranfield Demon in flight.

A parallel development programme by part-time Masters students at Cranfield University and BAE SYSTEMS produced the Kestral BWB demonstrator, which flew in 2003 (Fig. 4.23).

The Demon UAS technology demonstrator aircraft was developed as part of the BAE SYSTEMS/EPSRC funded programme of research for future UAVs. The author was chief engineer on the project, which incorporated technologies from several UK universities. The aircraft was designed and built at Cranfield University, with contributions from BAE SYSTEMS, and several subcontractor organisations.

Figure 4.24 shows an early CAD model, whilst Fig. 4.25 shows a flight test. The aircraft appears in the *Guinness Book of Records* as the first aircraft to demonstrate the use of fluidic flight controls.

CHAPTER 5
WHAT'S UNDER THE SKIN?: STRUCTURE AND PROPULSION

5.1 GENERAL

Having looked at the reasons for designing a new aircraft, its requirements and the reasons for its external shape, it is now the time to look at the interior. One might use the analogy of a human body to see that for successful operation it requires a skeleton, muscles, digestive system, sensors and a control system. These all work together efficiently even when they have different functions. The aircraft designer must produce a similar harmony within the aircraft interior. Each system is important, but each may have conflicting requirements. A good designer must weigh up the conflicts, use relevant analyses and synthesize the aircraft into an efficient whole. It is a truism that the outside of the aircraft has to be bigger than the space required inside the aircraft. It is often necessary to modify the external shape to accommodate the interior, with consequent aerodynamic changes.

To return to the body analogy, this chapter will describe the aircraft's skeleton, muscles, sensors, etc. in terms of structure and propulsion. Fuel, flying controls, avionics, furnishing and weapon systems will be described in Chapters 6 and 7.

5.2 THE STRUCTURE

An aircraft structure should be designed to meet a number of conflicting requirements which include:

   (i) Low weight.
   (ii) Acceptable material and manufacturing costs.
   (iii) Adequate strength to meet the maximum expected loads, with a suitable safety factor.
   (iv) Adequate stiffness so that distortions are kept within acceptable limits.
   (v) Good in-service properties, such as fatigue and corrosion resistance, together with tolerance of expected temperatures and other atmospheric conditions.

5.2.1 Materials

Aircraft designers have a wide range of structural materials to choose from, including those described in the following subsections.

5.2.1.1 Light Alloys

Light alloys are essentially based on aluminium and are used on the majority of aircraft. Their advantages are the high strength and stiffness-to-weight ratios, relatively low cost, general ease of handling, except in special cases, availability, choice of many fabrication processes and, above-all, familiarity. Against this is the dependence upon relatively low-temperature heat-treatment processes to obtain the desired material properties, which implies poor strength and stiffness at elevated temperatures and relatively poor fatigue properties. The higher-strength alloys are poorer in fatigue and their use is restricted to predominantly compression-sensitive areas. Tensile loads produce more fatigue problems.

The maximum temperature at which light alloys can be used for continuous operation is about 130 °C, and this is equivalent to a Mach number of about 2.2 due to kinetic heating, when allowance is made for heat dissipation by the structure. For short flight times, as on missiles, the maximum flight Mach number can be of the order of 3.0. Aircraft that do not use light alloys as the prime structural material are rare.

5.2.1.2 Titanium

Titanium has established its place in certain aircraft applications. It has some advantages, but also some disadvantages, not least of which is cost. Although titanium alloys have greater density than light alloy, they have a higher strength-to-weight and stiffness-to-weight ratio, particularly at elevated temperatures. Very close control is required to obtain consistent properties, and titanium is generally more difficult to handle in the process of manufacture.

Titanium is used when heat-resistant properties are required and when the local geometry is such that a denser material than light alloys is advantageous. Airframe examples are firewalls, main frames and very heavily loaded local structure. Areas close to engine exhausts and gun barrels often use titanium alloys.

Recent developments in super-plastic forming can give cost savings for the completed object. This process uses special heat treatment to form very complex shapes. Main savings are from the reduced parts count.

Titanium thermal expansion rates are similar to those of carbon fibre composites (CFC) and as it does not have the CFC/aluminium alloy corrosion problems, it may be successfully used in close proximity to CFCs.

5.2.1.3 Steel

Steel has always been used in airframes, and indeed the first metal aircraft used steel rather than light alloys due to the relatively high cost and poor availability of the latter at that time. The severe limitations of the use of light alloys at elevated temperatures, together with the problems associated with the use of titanium, forced designers to reconsider the design of structures in high-grade steels. The main problem was that of obtaining a light structure since, although steels have strength-to-weight and stiffness-to-weight ratios comparable with light alloys, the density is nearly three times as great. This implies minimum gauge design in many places and accentuated instability problems.

Small quantities of steel have always been used in airframes where high strength is required in a small space. Machining has presented some problems, but these have now been overcome and supplemented by the development of suitable chemical etching processes. Highly stressed components, such as landing gear and engine pylons, make extensive use of steel.

The cases of many earlier solid rocket motors were welded steel units since, although titanium is lighter, the cost was prohibitive on an expendable vehicle. Some large liquid rocket vehicles also use thin steel construction, but it appears that light alloy is adequate for this purpose.

5.2.1.4 Magnesium Alloys

Magnesium alloys are very attractive for many applications. The main use is for large castings where bulk is essential and the very low density of magnesium results in a weight saving. There are certain difficulties, which include the low resistance to corrosion and difficulty of making joints without welding due to crack propagation. Perhaps the biggest drawback is unfortunate past experiences with corrosion of the material. The cost is not very high, however.

5.2.1.5 Non-Metallic Materials

The non-metallic group of materials covers a wide range, the most important of which, from the point of view of aircraft designers, are reinforced plastics, rubber, sealants and cockpit transparencies. Wood has been shown to be an attractive material for some small, home-built aircraft. Modern adhesives give better service life than older glues. Some small aircraft have been constructed using reinforced plastics, such as resin-bonded glass fibre, but then use has been restricted primarily by the poor stiffness qualities of glass fibre. There is a temperature limitation by virtue of the resin used. Application of reinforced plastics in mixed structures has been limited by the difficulty of attachment to metal parts. However the advent of carbon fibres has changed the overall picture and large primary components are in service in these materials. The main problems being overcome are service behaviour and, often, labour-intensive production.

Figure 5.1 shows the Beech Starship, a business aircraft, which has a largely composite construction. Wing skins and fuselage panels are constructed from carbon-fibre honeycomb panels, which give good stiffness and minimize the number of ribs and frames.

5.2.2 Load-Carrying Methods

A structure is a device for transferring mechanical loads from one point to another using the following mechanisms (Fig. 5.2).

Fig. 5.1 The Beech Starship.

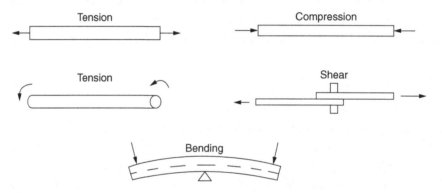

Fig. 5.2 Structural load transfer mechanisms.

5.2.2.1 *Tension*

The simplest way of carrying a load is to use a tension member. Such a member is inherently stable and it gives the lightest solution. However, a point of importance with regard to tension members is that a failure implies severance of the structure, normally with catastrophic results, unless alternative load paths are provided.

5.2.2.2 *Compression*

An ideal member in pure compression is stable in the same way that a tension member is. However since the ideal is never achieved in practice, all compression members must be designed with consideration of the question of instability. Extra material is therefore required relative to that necessary to take the simple direct load in compression. There are three basic forms of compression instability:

(i) Overall, which is a simple strut failure (sometimes called Euler buckling).

(ii) Local, where a flange or other part of the local cross-section buckles in a short distance.

(iii) Torsion, where the whole section twists, but it is frequently difficult to distinguish this from local instability.

A compression member is often capable of transmitting some load after initial buckling has occurred. In some cases this characteristic is used in the design.

5.2.2.3 Shear

Although shear is a basic form of loading, the mechanism by which a shear member reacts to load can be considered as a combination of diagonal tension and compression.

Since the shear member can be made to carry a large portion of the load in tension it is quite an efficient unit. The presence of buckles prevents the use of a tension field member on aerodynamic surfaces, up to the maximum normally expected load, because the buckles distort the shape and adversely affect the aerodynamic flow.

5.2.2.4 Bending

In reality a beam is not a single type of load-carrying member, but a simple structure. The loads are carried by a combination of tension, compression and shear.

Under load, a beam will curve and the inside edge is in compression. The presence of the shear-carrying centre, or web, helps to stabilize the compression surface. The strength of a beam varies according to the shape of the cross-section. If this is rectangular, the strength is proportional to the square of the depth, but in the case of a T section it is more nearly directly proportional to the depth.

In a very deep beam the edge conditions become secondary and a shear member is left. The use of a beam is inevitable when loads must be transferred through a distance normal to the line of action of the load, and hence are very common in aircraft structures.

5.2.2.5 Torsion

Torsion is really a special form of shear, and the same remarks apply in general. Torsion usually occurs in structures when the load path changes direction, such as lateral loads on the fin, twisting the rear fuselage.

Appendix A gives formulae that may be used for simple structural elements, subject to the above loading methods.

5.2.3 The Main Structural Members

5.2.3.1 Flying Surfaces

The flying surfaces, as typified by the wing, are essentially beams. As such they are far from ideal, since the depth is not great. A reduction in the thickness/chord ratio is critical since the bending strength is proportional to this dimension.

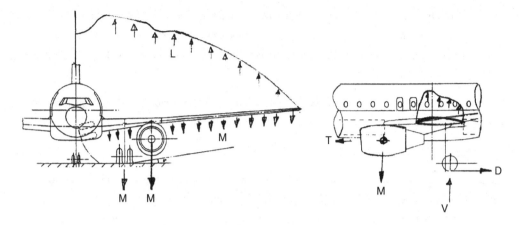

Fig. 5.3 Major wing structural loads (M = mass, L = lift, T = thrust, D = drag (landing gear),
V = vertical force (landing gear)).

In addition to acting as a beam to resist spanwise airloads, the wing must also carry considerable torsion loads and provide sufficient stiffness to prevent excessive twist. The torsion load arises primarily from flaps and control surfaces, and the torsional stiffness is necessary to prevent divergence, flutter and control reversal. A reduction of thickness/chord ratio directly reduces the enclosed area, and thus the effect on twist is varied as the square of the depth. Figure 5.3 shows the spanwise and chordwise loads.

Reduced aspect ratio reduces both bending moment and twist and this is thus structurally advantageous. Increased taper has the effect of moving the airload towards the root, thus reducing the bending moment. Sweepback not only effectively increases the length of the beam, but also tends to cause the airload to shift outboard, and is thus a disadvantage. In addition large local loads can occur at the rear of the wing around the root. Recent advances in manoeuvre load control (MLC) and gust load alleviation (GLA) actively modify the airload distribution.

In GLA, the flight control system senses an up-coming symmetric gust and sends a signal to the ailerons. These rapidly deflect symmetrically to reduce wing-tip camber. This reduces wing-tip lift and the spanwise centre of pressure moves inboard. The lift reduction reduces wing shear force, and also reduces bending moment. The Airbus Industries A320 used this system to reduce the fatigue damage to the wing, thus leading to reduced material and weight.

It can be seen from the above that the wing has two main functions, as a beam and as a torsion box. These functions can be separated by having separate end-load carrying booms and shear skins. A more efficient solution is to combine the role of the skins, stabilizing them with long struts, or stringers. When the skin is very thick, the stringers may be dispensed with. Often two webs are insufficient to carry the vertical shear load and as the wing is thin, the stringers will tend

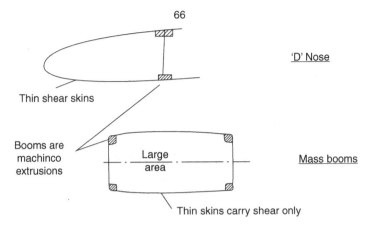

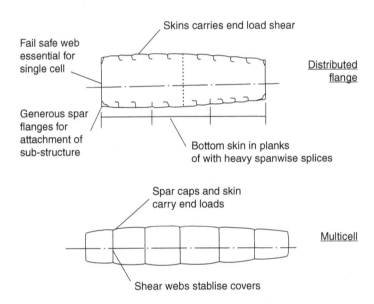

Fig. 5.4 Alternative forms of wing-box construction.

to touch in the middle. A multi-cell box results. Figure 5.4 shows different constructions, whilst Fig. 5.5 shows how the wing components fit together.

The other main components shown are wing ribs, which have several functions. They support the skin to form the aerofoil shape, and stabilize the stringers to stop them buckling. Flaps, ailerons and engines introduce concentrated loads into the wing, which must be carried by heavier ribs. Most aircraft use the area between the spars as a fuel tank. Special ribs are used to seal the tanks (tank-end ribs) and some intermediate baffle ribs may be used to prevent fuel sloshing.

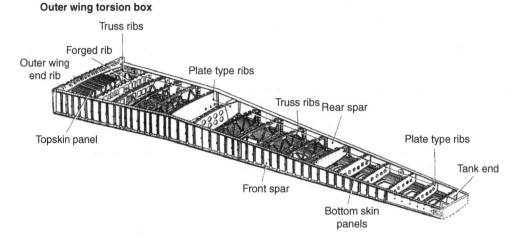

Fig. 5.5 The wing-box configuration of the F-100 airliner.
(Source: Fokker.)

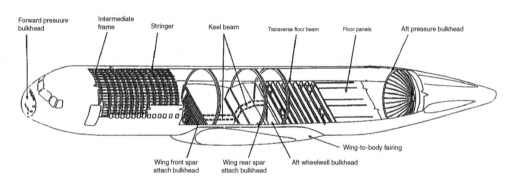

Fig. 5.6 Major fuselage structural member types.

5.2.3.2 Fuselage

The loads carried by the fuselage are basically similar to those imposed upon a wing. Bending can take place in either a vertical or a horizontal plane, due to tailplane and fin loads, respectively. Fin loads also impose a torque, as do asymmetric tailplane loads. Undercarriage loads impose vertical bending shear and torque loads, which may be critical near the nose.

Unlike a wing, however, the fuselage normally has ample depth for bending and an enclosed area for torque, and is thus a more nearly 'ideal' type of structure. Fuselage construction is usually one of two forms. It can be either a fully effective beam with stringers stabilizing the skins (Fig. 5.6) or alternatively a number of longitudinal, discrete members with shear-carrying skins can be used. These members are known as longerons. When the top and bottom portions of the structure are badly cut away, they may be used as fairings only and one or two 'floors' or

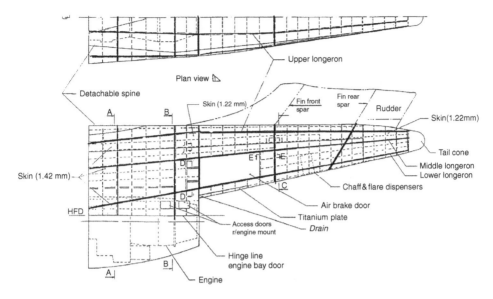

Fig. 5.7 Rear fuselage-longeron and shell construction.

'decks' are introduced to complete the torsion box. Cross-sectional frames give section shape, transmit local loads to be shell, and supply edge supports for the stringers and longerons (Fig. 5.7).

Pressure cabins are normally integral to the fuselage structure. Passenger and freight floors usually contribute to the resisting of bending loads, as well as transmitting local loads to the shell. The main reason for this is the difficulty of isolating them from the main beam.

Wing bending loads are sometimes transmitted round the fuselage frames, but where possible, taking the loads directly across the fuselage leads to a lighter structure.

5.3 PROPULSION: THE PRIMARY POWER SYSTEM

All aircraft, apart from balloons or gliders, need a propulsion system for sustained flight. A vast array of engines is on offer, but consideration of the aircraft's required flight speed limits the options available.

An obvious distinction can be made between air-breathing engines and rockets. At least in terrestrial applications the former have a relatively high installed engine weight, but low fuel consumption, while the opposite is true of the latter. Thus air-breathing engines are used where long-time operation is required and rockets where the operation can be completed in a short time or where the atmosphere is too thin to support combustion. Figures 5.8 and 5.9 summarize some of the characteristics of powerplants.

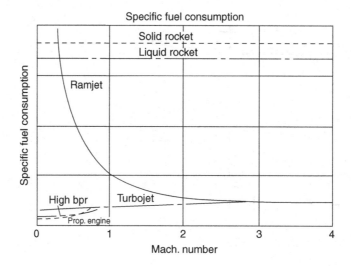

Fig. 5.8 Trends of specific fuel consumption with Mach number (BPR = bypass ratio).

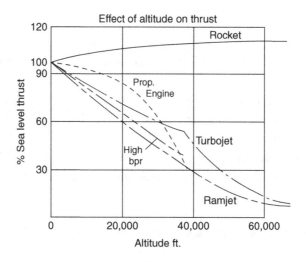

Fig. 5.9 The effect of altitude on thrust.

5.3.1 Air-Breathing Engines

5.3.1.1 Piston Engine, Driving a Propeller

Apart from small, light aircraft where initial cost is of prime importance, the piston engine has few applications in aeronautics. It is unlikely to be used in aircraft flying faster than M 0.3. The installed weight for a given power is relatively high.

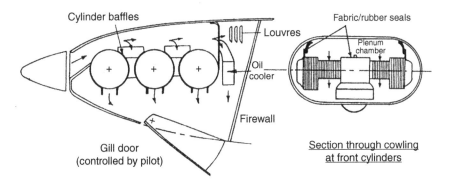

Fig. 5.10 Light aircraft air-cooled engine.

An important consideration with piston engines is the method of cooling. The vast majority of current engines are air-cooled, as shown in Fig. 5.10. This is a very simple system, but it does involve a considerable drag penalty, which is not too critical in light aircraft. A more complicated alternative is the liquid-cooled engine, which reached its development peak in World War II fighters and bombers. This produces a very clean engine installation, but the radiator that is used to cool the liquid creates drag.

5.3.1.2 Turbo-Prop and Prop-Fan Engines
The turbo-prop engine is the most suitable powerplant for subsonic speeds up to about M 0.65. A turbine engine drives a propeller through a reduction gearbox. It has lighter weight than a piston engine, despite the reduction gear. It has low fuel consumption, smooth running and secondary advantages such as a readily available source of compressed air. The usual speed regime is actually M 0.3 to M 0.65 with a tendency to the higher figure on longer range aircraft.

Figure 5.11 shows a turbo-prop engine installation from the Cranfield F-93B project. The early 1980s saw a sharp rise in fuel prices and this led to a new variety of turbo-prop engine. Variously called prop-fan, unducted fan or open rotor, such engines have a conventional turbine core driving advanced propellers. They use supercritical section blades with swept back tips. Most examples have two rows of contra-rotating blades with five or six blades per row. They have been designed to cruise at high efficiency at M 0.8 and have a claimed fuel consumption reduction of some 30% relative to current turbo fans. Blade-tip noise, vibration and possible blade shedding usually leads to the choice of aft fuselage engine mounting as on the A85 project with Rolls Royce RB-509 project engines (Fig. 5.12).

5.3.1.3 Turbo-Jet Engine
Above about M 0.65, blade compressibility effects cause a reduction in the efficiency of conventional propeller engines and reaction propulsion is used. The turbo-jet engine is most satisfactory

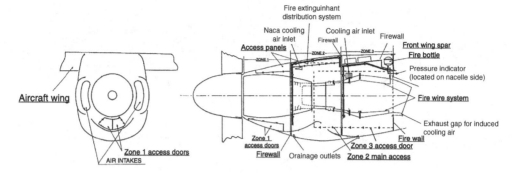

Fig. 5.11 Turbo-prop engine.

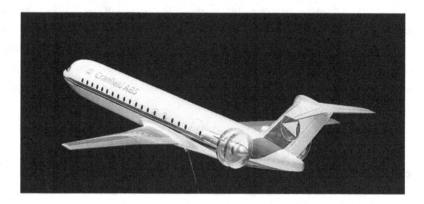

Fig. 5.12 Cranfield A-85 with prop-fan engines.

for use in the M 0.75 to M 3.0 speed range at altitudes up to about 18 km, although in fact for aerodynamic reasons few aircraft operate between M 0.9 and M 1.4.

Fuel consumption and noise are high on the turbo-jet relative to the turbo-fan and thus few commercial aircraft currently use them.

5.3.1.4 Turbo-Fan or Bypass Engines

Figure 5.13 shows a high bypass-ratio engine which bypasses some of the compressed air round the hot section of the engine. These engines use a large fan to provide much of the thrust and in some cases this implies a three-shaft engine. Bypass has the advantage of reducing both fuel consumption and noise, and taken to the extreme may use a geared, variable-pitch fan, which effectively makes the engine a ducted turbo-prop. Of course these more sophisticated engines are heavier than simple turbo-jets with a single rotating assembly and no bypass.

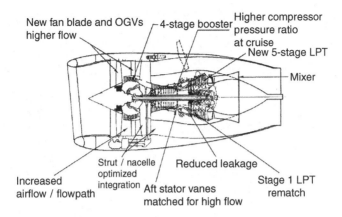

Fig. 5.13 Bypass (turbo-fan) engine (highlighting changes from an earlier model).

5.3.1.5 Reheat Turbo-Jet Engine

The addition of a second combustion stage to the turbo-jet engine, by injection of fuel into the exhaust, extends the useful speed regime up to M 3.0. Reheat can be used in two ways, either as a means of increasing thrust quite considerably for short periods of flight, e.g. the transonic phase, or for continued operation in the M 2.0 to M 3.0 regime.

Fuel consumption is dramatically increased, but a smaller core engine may be used for the whole flight than if a larger engine were designed to cater for a short-term high thrust requirement. Re-heat is also called after-burning.

Burning extra fuel in the cold airstream of a bypass engine appears to be a profitable way of increasing thrust, especially when the vehicle requires a large range of thrust for different flight regimes.

5.3.1.6 Ramjet

The ramjet is a simple, lightweight engine which basically dispenses with all moving parts. Since its operation relies upon the compression effect of forward speed and expansion through a nozzle the engine is not really suitable for flight below about M 2.0. The upper limit is around M 7.0. When used on aircraft, a second powerplant for low-speed flight is necessary, and hence little application has been made in this connection. However, the ramjet is ideal for certain missiles, particularly those flying over relatively long range at altitudes below about 22 km.

New space-launcher concepts have led to the project design of turbo ramjets. The core of the engine is a turbo-jet that propels the aircraft from static conditions to M 2.0 or 3.0. At this speed the annular ramjet chamber is used and the turbine closed down. The lower stage of the two-stage SL-86 project would use turbo ramjets fuelled by liquid hydrogen up to separation of both stages at M 4, 25 km altitude (Fig. 5.14). The upper stage would fly into low Earth orbit, powered by a liquid rocket. Details of the turbo-ramjet engine are shown in Appendix A.

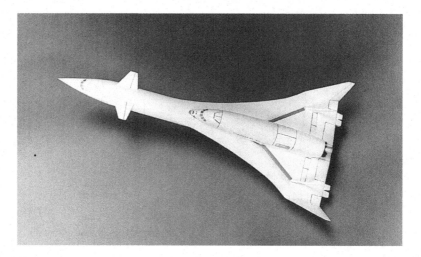

Fig. 5.14 Cranfleld SL-86.

5.3.2 Rockets

Rockets are more efficient than air-breathing engines at speeds above about M 3.0. Although propellant consumption is inevitably high, due to the need to carry oxidant as well as fuel, in this lies one of their great advantages of independence of altitude. Chemical rockets are normally classified into two types.

5.3.2.1 Solid Fuel Rockets

The solid rocket is an explosive whose burning is carefully controlled by design. Burning usually takes place radially rather than longitudinally, as this results in a more efficient motor. Hence, thrust is a function of length, and burning time of diameter. The charge is 'star'-shaped internally to give a uniform burning periphery and hence uniform thrust. Installation is simple, handling easy and reliability high. The unit is also cheap, at least in smaller sizes. Thrust cannot be controlled easily, although it is possible to control thrust direction by exhaust vanes or nozzle gimballing.

There are three fields of application of the solid rocket:

(i)   As motors for small cheap missiles of relatively short range.

(ii) Large ballistic missiles where the ease of handling and reliability offsets its poorer performance relative to that of the liquid rocket.

(iii) Assisted take-off of manned aircraft, and boosted missiles, such as the 'Sea Dart'. The rockets boost the missile to a speed sufficient for the operation of the sustainer ramjet.

5.3.2.2 Liquid Fuel Rockets

In large sizes the liquid fuel rocket has a lighter installed weight than its solid counterpart, and also a lower propellant consumption. Thrust and thrust direction are controllable, but cost is higher and reliability less than for a solid unit. This is due to the complexity of the fuel system.

At present the liquid rocket is used to power space vehicles, large ballistic vehicles and manned aircraft that fly at high altitudes or require a high thrust-to-weight ratio at medium altitudes.

5.3.3 Location of Powerplants

5.3.3.1 Installation Clearances

The powerplant must be located to ensure that there is adequate ground clearance during taxi-ing, take-off and landing. In many cases there are also restrictions imposed to ensure that there is adequate airframe clearance.

This requirement normally fixes the position of propeller engines, either in the nose of the aircraft or in wing nacelles. A few single-engine layouts use a pusher propeller with either a long drive shaft or twin booms. Variation in the vertical and spanwise location of wing nacelles may be possible to some extent and on short take-off and landing (STOL) designs an attempt is often made to spread the beneficial slipstream effects over as much of the wing span as is possible. It should be noted that this effectively increases the speed over the wing, and hence the drag, in the cruise condition. Figure 5.15 shows a number of single- and twin-engined propeller installations.

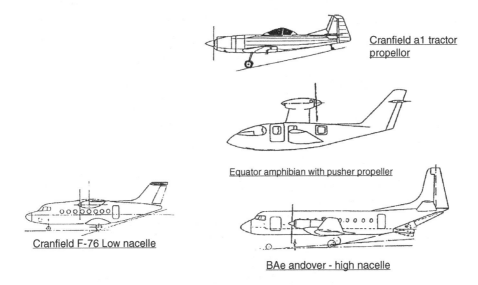

Cranfield a1 tractor propellor

Equator amphibian with pusher propeller

Cranfield F-76 Low nacelle

BAe andover - high nacelle

Fig. 5.15 Single- and twin-propellor installations.

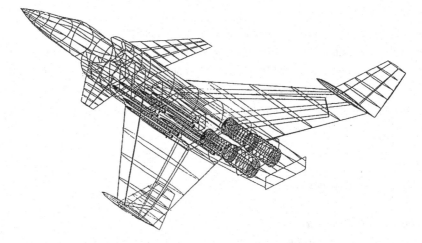

Fig. 5.16 Cranfield TF-89 with fuselage-buried engines.

Pod-mounted jet or turbo-fan engines that are located below the wing of an aircraft may be restricted in position by the need to provide adequate ground clearances. The critical case is landing with, say 5° of bank and it is greatly aggravated when the wing is swept back. A further important point with regard to low-mounted air-breathing engines is the need to eliminate, as far as is feasible, the possibility of stones and other debris being drawn into the air intake. In particular, the intakes should not be located immediately behind wheels, as even when mudguards are provided, there is a serious danger of debris ingestion. These difficulties are aggravated with increase of bypass ratio and reduction of aircraft size.

Early jet transports had engines buried in the wing and most current military aircraft have engines mounted in the fuselage (Fig. 5.16). These installations have reduced drag, but can lead to structural complexity and are inflexible should a larger size engine be required later.

5.3.3.2 Overall Propulsive Efficiency

The overall propulsive efficiency is determined by the individual efficiencies of the air intake and exhaust nozzle, or propeller design. In the latter case it is feasible to expect the propeller efficiency to be of the order of 90% under cruising conditions, but somewhat less at lower speeds. When good take-off performance is required it may be necessary to compromise.

Intake efficiency is critically dependent upon Mach number. Up to a Mach number of around 1.6 a simple pitot intake is adequate and it should be as short as is feasible. At higher Mach numbers more complex arrangements are necessary to deal with the shock waves present, and variable geometry is desirable to maintain high efficiency over the whole flight speed range, as is shown in Fig. 5.17.

In these circumstances the intake duct must inevitably be long. The location of the intake has an important effect upon its efficiency. Nose intakes of minimum necessary length are

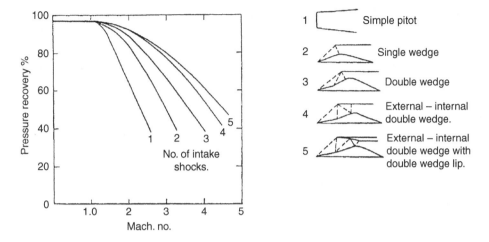

Fig. 5.17 Supersonic intake efficiency.

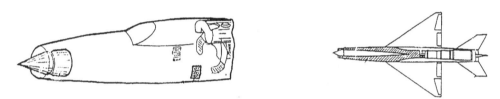

Fig. 5.18 Supersonic nose intake, MiG-21 (note auxiliary intakes below cockpit).

best (Fig. 5.18). Intakes mounted alongside a structure give rise to boundary layer and shock problems, although the latter may be favourable in certain circumstances.

Intakes located below a wing or fuselage usually have a higher efficiency than those mounted above. Circular intakes are easier to manufacture within a given tolerance than square ones, but the incorporation of variable geometry is more difficult unless they are axi-symmetric or semi-axi-symmetric.

The exhaust system should be as short as is possible, to minimize noise, vibration and heat problems on the aircraft structure. Fixed-geometry nozzles are satisfactory for use up to low supersonic Mach numbers. Convergent, or preferably, convergent-divergent, nozzles become increasingly important at higher Mach numbers due to the need to increase exhaust velocity.

5.3.3.3 Other Factors

Engine location has a marked effect on stability and control, with wing-mounted engines giving a more flexible longitudinal CG range. Wing-mounted engines, however, often drive the sizing requirements of the fin and rudders due to single engine failure on multi-engined aircraft. Figure 5.19 shows a powerplant installation comparison between two 150-passenger aircraft. Table 5.1 shows some of the strengths and weaknesses of both configurations.

Table 5.1   Comparison between wing and fuselage-mounted jet engine pods

Criterion	Wing-mounted	Fuselage-mounted
Ground clearance	Possible problem (particularly smaller)	Good
Internal noise	Fair	Good
Acoustic fatigue	Possible flap and wing problem	Probable fuselage problem
Crash safety	Good	Possible problem
Propulsive efficiency	Good	OK if well positioned
Longitudinal stability	Good – delays tip stall	Loading problems short tail arm and tip stall
Asymmetric thrust	Poor	Good
Weight	Good – wing bending and torsion relief	Poor – heavy tail, heavy fuselage frames, 2–4% heavier empty
Maintenance	Usually good	Often high off ground
Wing aerodynamic efficiency	Possible flap and leading edge cutouts	Very good
Fuel feed to engines	Good	Fuel lines pass through cabin
Anti-ice	Easy to use hot air on wings	Ducts pass through cabin

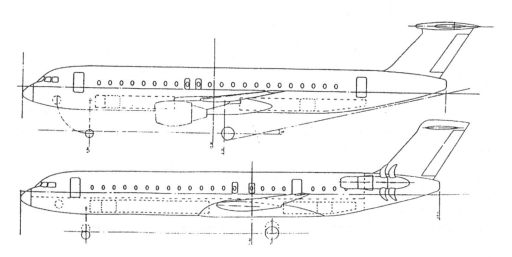

Fig. 5.19 Propulsion system comparisons wing- and fuselage-mounted engines.

Aircraft engines can have a marked effect on the world environment. There are obvious effects from engine noise, and pollution from the exhaust emissions. Less obvious is possible ozone-layer depletion due to emissions of gases near the tropopause. Airworthiness requirements are imposing ever more stringent limits to control these effects.

Chapter 6   What's Under the Skin?: Airframe Systems

6.1 Secondary Power Systems

We have looked at an aircraft's primary power system, its propulsion, and will now examine its secondary power systems (SPS). Secondary power is usually provided by an aircraft's main engines when they are operating, but can be supplied by other sources such as an auxiliary power unit (APU) in the aircraft or an external source on the ground.

Almost all present aircraft SPS are of three types, shown in Fig. 6.1.

Power is extracted from turbo-fan powerplants in two forms. Bleed air is tapped from one or more points along the engine compressor to provide pneumatic power. Drive shafts, from the engine's high-pressure shaft, drive an accessory gear box on which are mounted generators and hydraulic pumps, as well as the engine's own accessories. A typical large turbo-fan is shown in Fig. 5.13.

Turbo-jet powerplants can also supply bleed air and shaft power for systems, but turboprop powerplants in general have difficulty in supplying sufficient bleed air for airframe requirements. In such a case, extraction of shaft power to drive separate compressors may be considered.

SPSs are often linked with each other, but it is convenient to separate them into the types described below.

6.1.1 Air-Conditioning and Pressurization

On many civil transports the largest amounts of secondary power are extracted in the form of bleed air and the largest continual user of this power is the aircraft's air-conditioning and pressurization system.

It is usual to maintain an aircraft cabin at a pressure equivalent to that at 8000 ft or less, and this requires a constant supply of pressurized air. In addition to this, oxygen used by the occupants must be replaced and the temperature and contaminant levels controlled, requiring a significant throughput of air, although some of it is recycled through the cabin, after filtering.

It is the control of temperature that usually determines the rate at which air must be supplied and, with the type of air-conditioning system installed on most aircraft, this also requires the air be supplied at pressures significantly above those in the cabin.

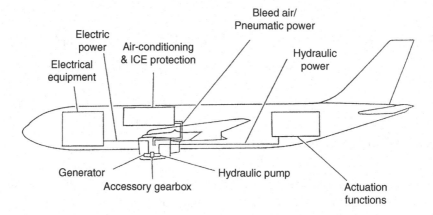

Fig. 6.1 Secondary power systems.

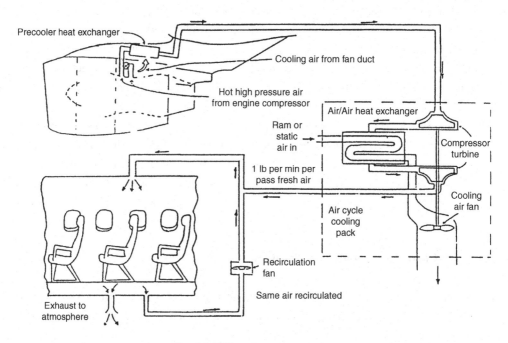

Fig. 6.2 ECS system components.

Figure 6.2 shows the major elements of an air-conditioning system. There are extremely remote events where the pressurization system fails. In this case, on-board emergency oxygen systems are used to preserve life. The combination of air-conditioning, pressurization and oxygen systems are usually termed the environmental control system (ECS).

In addition to cooling the crew and passenger compartments on high-speed aircraft, it may be necessary to remove heat from the undercarriage bays, electrical equipment and flying control areas.

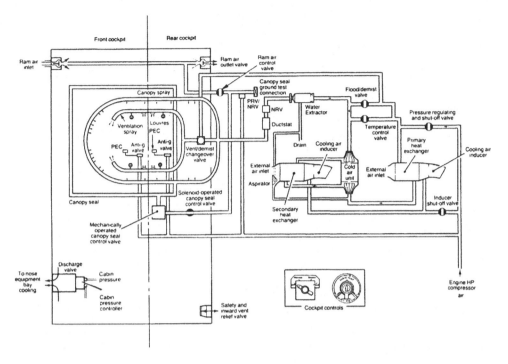

Fig. 6.3 Hawk ECS system.

Military aircraft use ECS systems working on the same principles as those for civil aircraft, but with the following changes:

(i) The pressurized cabin has a much smaller volume and works at a lower pressure differential.

(ii) Oxygen systems are used in normal operations at higher altitudes.

(iii) Military avionics are more complex and produce comparatively more heat, and thus require more cooling.

The BAE SYSTEMS Hawk trainer system is shown in Fig. 6.3.

6.1.2 Ice Protection Systems

There have been a number of accidents resulting from the accretion of ice on the lifting surfaces of aircraft. Dangerous icing conditions are rarely encountered by modern aircraft, but it is normally necessary to provide protection. Engine intake icing is more common due to the special flow conditions present and most aircraft at least have a provision for de-icing in this region of the

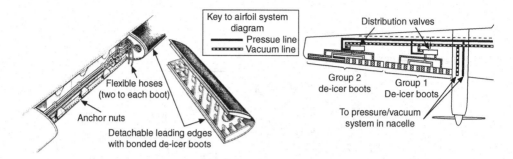

Fig. 6.4 Rubber 'boot' de-icing.

airframe. Protection against icing can be obtained in various ways. De-icing allows ice to form to a certain extent and then periodically removes it. Anti-icing is a system that is activated when icing conditions are likely and prevents ice formation.

Bleed air is not only at high pressure, but also high temperature and thus on turbo-fan-powered aircraft it usually provides the main form of ice protection for the airframe and engines.

On large transports, only the outer portion of the wing leading edges or slats and the engine cowl lips require protection, but the air-flow requirement can be similar to that for air conditioning. Smaller aircraft, such as the BAE SYSTEMS 146, also require protection of the tail surface leading edges, and when operating ice protection can take three times the flow for air-conditioning, for the short time that ice protection is required.

Turbo-prop powered aircraft have insufficient bleed for thermal anti-icing, so alternative forms of protection are required. Some leading edges are protected using a pneumatic powered system, but this uses only small amounts of air to periodically inflate flexible 'boots', thus de-icing the area. These 'boots' do use much less power than thermal anti-icing, but have disadvantages, in that they are not totally effective in all conditions and are prone to erosion due to their relatively soft nature (Fig. 6.4).

Another possibility is to spray a fluid over the affected surface to reduce the melting point of the mixture formed to below the ambient temperature. Localized anti-icing may be provided by the use of embedded electrical heating elements.

A more recent development has been electro-induction de-icing. When ice is detected, the system uses electro-magnetic means to vibrate the the wing leading-edge to shake ice off.

6.1.3 Electric Power Systems

On medium and large aircraft, electric power is distributed in two forms. Engine-driven generators provide three-phase 400 Hz 115 V line to neutral AC power, but some of this power will be converted to 28 VDC. This latter form may be the only one used on smaller aircraft.

Most aircraft, however, have both direct and alternating current supplies, with the former used primarily for such items as electrically driven instruments and lights. AC sources provide power for high-energy devices such as electrical anti-icing. The usual sources of power supply are engine-driven alternators, which supply AC that can also be converted to DC by means of transformer/rectifier units. Batteries are also used to supply DC directly and can be used to start engines on small aircraft. Larger aircraft use small turbines as APUs for engine starting and air conditioning whilst the aircraft is on the ground. Some twin-engined aircraft used for long over-water flights use flight-rated APUs as a further electrical power back-up.

Power supplies must be duplicated or triplicated so that power can be suplied even in the event of failures of individual components. Figure 6.5 shows a schematic of an airliner system, whilst Fig. 6.6 describes the BAE SYSTEMS Hawk's system.

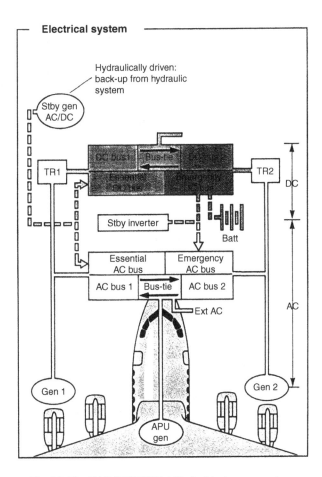

Fig. 6.5 The BAE SYSTEMS 146 electrical power systems.

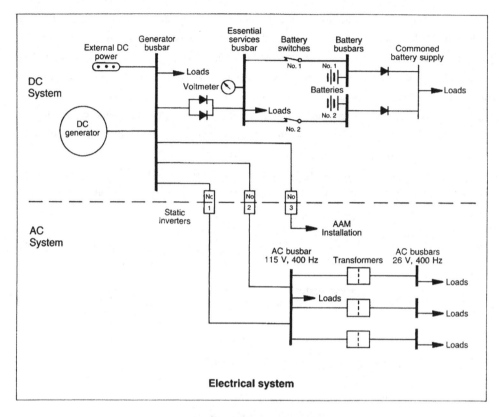

Fig. 6.6 The Hawk electrical power system.

6.1.4 Hydraulic Systems

The major use of hydraulics on aircraft is to provide for the actuation of devices such as flying controls. Functions actuated can vary greatly between aircraft, and on the same aircraft, but almost all medium and large civil transports have a system operating at a pressure of 3000 psi.

The other main need for hydraulic power is the undercarriage. It is used for retraction jacks and for braking. Large doors are often actuated hydraulically High reliability is an important consideration and this leads to the duplication or triplication of hydraulic systems, as in electrical systems. This is particularly important in flying control actuation, as shown in Fig. 6.7 for the BAE SYSTEMS 146 and Fig. 6.8 for the Hawk.

On a large transport aircraft with three hydraulic channels, the mass of hydraulic pipework and fluid contained within it is considerable. Increasing the pressure allows reduced flows of fluid and thus pipe size. Actuator sizes can also be reduced, but limits are imposed by the higher pressures to be contained by pipe walls. Reductions of 25% in hydraulic system mass have been claimed for 8000 psi systems relative to present standards. However, care must be taken to examine the assumptions on which these claims are based.

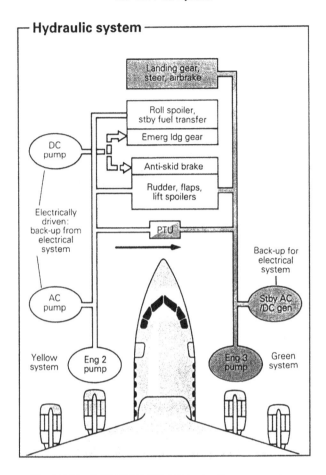

Fig. 6.7 The BAE SYSTEMS 146 hydraulic power system.

Modern civil and military aircraft with 'active' controls (see paragraph below) require very rapid flying control actuation. This is usually supplied by hydraulic actuators, which require specially designed hydraulic systems that will provide for the very high fluid flow rates required.

6.2 THE FUEL SYSTEM

On large aircraft with several engines and numerous fuel tanks, the fuel system can be exceedingly complicated. A large number of design requirements must be met as a fuel system failure may result in a crash. Probably more serious accidents occur to aircraft in the early development stages due to faulty fuel systems than through any other cause. In spite of the complexity of the requirements it is always desirable to make the system as simple as possible.

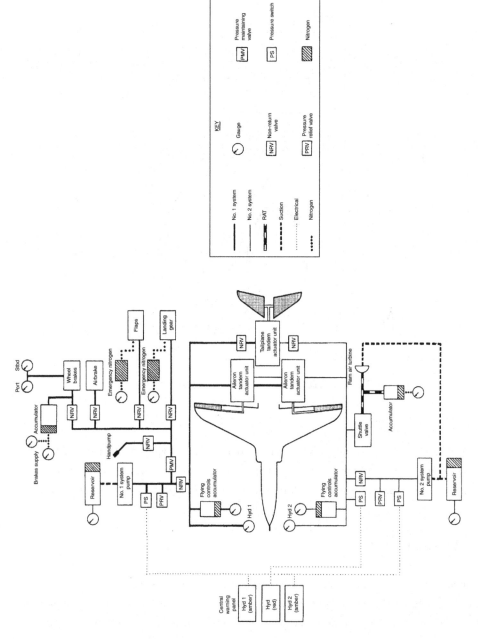

Fig. 6.8 The Hawk hydraulic system.

Fuel system schematic

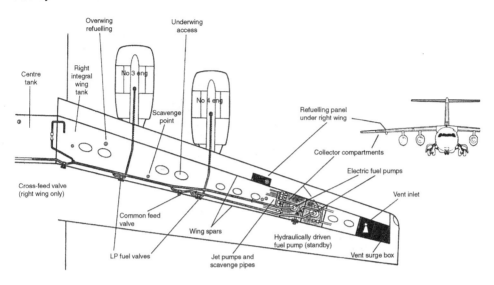

Fig. 6.9 Fuel System of the BAE SYSTEMS 146.

Safety risks come from fuel starvation, which leads to engine failure or fire due to uncontrolled fuel ignition. Fuel system design must prevent either of these failures and must produce a system with high reliability and maintainability.

It is often a requirement that fuel can be jettisoned or taken on in flight. Numerous valves, pumps and interconnections are necessary, each one increasing the possibility of unreliability. A failure of these components must not affect the performance of the aircraft, so that duplication and hence more complexity is necessary. It is usual to make full-scale models of the fuel system and test them under as representative conditions as can be obtained on the ground.

Most civil aircraft store fuel in their wings, as its weight acts in the opposite direction to wing lift, thus reducing wing bending moment and weight. This technique also has the advantage of isolating fuel away from the passenger compartment. Figure 6.9 shows a fuel system that is typical of subsonic airliners. Supersonic aircraft experience aerodynamic changes that affect the longitudinal trim of the aircraft. One way to combat this is to deflect control surfaces, but this causes extra aerodynamic drag. Figure 6.10 shows the ingenious method used by Concorde in which fuel is pumped from tanks 1 and 2 to 10 to alter the aircraft's CG.

Fuel systems of military aircraft lead to different problems, the main one being vulnerability to enemy fire. Analysis of Vietnam and Middle East wars showed that 62% of single-engined aircraft losses were attributable to fuel systems and tanks.

A number of measures are used to protect fuel systems. A simple method is to use kerosene fuel rather than the more flamable JP4.

The use of fuselage tanks will afford some protection by the fuselage structure. Another technique is to use self-sealing tanks and couplings. Valves and pumps can be placed within the protected tank and protected by armour. The air volume above the fuel may be pressurized with an

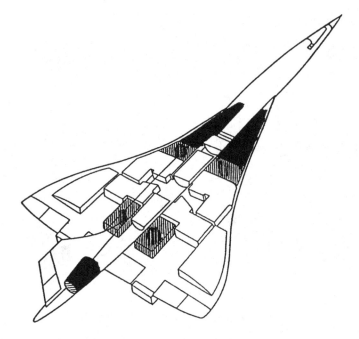

Fig. 6.10 Concorde's fuel system.

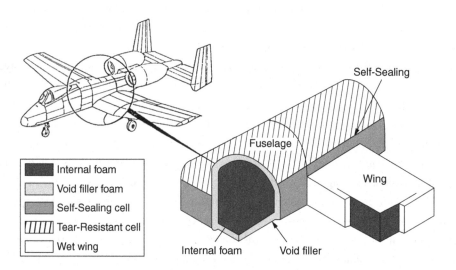

Fig. 6.11 The fuel system of the Fairchild A-10A.

inert gas such as nitrogen to prevent explosions. Another problem is leakage of fuel through holes. Self-sealing, above, helps, as does reticulated foam and sucked vent systems. The latter employs engine bleed air to drive jet pumps, which apply suction to the tank to prevent leakage. All of these techniques have their value, but their cost, weight, complexity, etc. must be assessed. The system used on the A-10A is shown in Fig. 6.11.

A340-200

16 sleeperettes + 42 Business + 181 Economy = 239 seats

Fig. 6.12 A340 seat layout.

6.3 FURNISHINGS

The furnishings are often determined by operator requirements, but the designer must make provision for them. Apart from seats, the items covered include removable bulkheads, wall trim, carpets, light luggage racks, lockers, galleys and toilets.

This aspect of aircraft design may seem trivial in comparison with other systems, but in civil aircraft design, furnishings have a marked effect on weight, passenger appeal, passenger safety, air conditioning, and electrical power and lighting.

Civil designs must be flexible enough to provide a wide variety of layouts ranging from high-density tourist seats through mixed class, executive layout to mixed passenger/freight arrangements. Figure 6.12 shows a typical civil aircraft layout. One important factor is the provision of parallel seat rails in the floor. These permit rapid relocations of seats, bulkheads, etc.

Furnishings on military aircraft are decidedly more spartan than those for civil aircraft. In combat aircraft they are limited to elector seats and a little trim.

6.4 SAFETY INSTALLATIONS

When an aircraft is designed to fly over wide expanses of water, provision must be made for such safety items as life jackets and dinghies. It is also necessary to incorporate emergency exit chutes on large aircraft to allow total evacuation within 90 seconds.

An important consideration in fuselage design is the provision of sufficient doors and emergency exits, and adequate access to them.

The principal item of military aircraft safety equipment is the ejector seat. Figure 6.13 shows a modern lightweight ejector seat developed for light trainer and ground attack aircraft. There is a powered harness retraction mechanism, which automatically pulls the pilot back into his seat for a safe ejection. The leg restraint lines also pull the legs into a safe position. The rocket motor is powerful enough to eject the pilot safely at zero forward speed and altitude.

The pilot usually sits under a transparent canopy, which must not block the seat's ejection path.

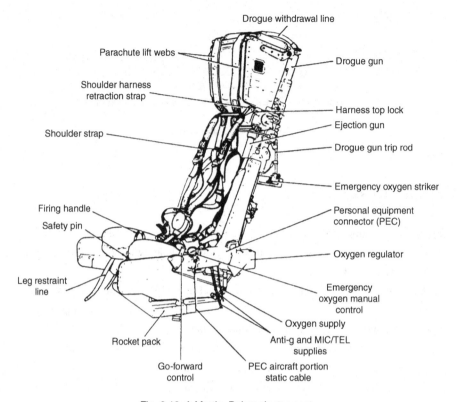

Fig. 6.13 A Martin–Baker ejector seat.

On initiating the ejection sequence, the miniature detonating cord (MDC), which is located around the edge and across the top of both canopies shatters the transparency to allow safe ejection. If this system fails the canopy may be broken by a strong piece of structure on the top of the ejector seat.

6.5 LANDING GEAR INSTALLATIONS

These form the vital interface between the ground and aircraft for the taxi, take-off and landing phases of operation. As in many aeronautical applications, a different terminology has been used on the opposite sides of the Atlantic ocean. The term undercarriage has been used in the UK, but landing gear is now used more universally.

6.5.1 Wheel Arrangements

Many types have been used in the past, but the three important arrangements are as follows.

6.5.1.1 Nosewheel Layout

The nosewheel, or tricycle layout, is the one that is normally used (Fig. 6.14). The advantages and disadvantages of this arrangement are largely opposite to those of the traditional tailwheel type. The major advantages are:

(i) While the aircraft is on the ground, the fuselage, and hence the cabin floor, is always roughly horizontal.

(ii) The view of the pilot when taxi-ing is relatively good.

(iii) The nosewheel acts as a prop to prevent overturning during braking.

(iv) The initial take-off attitude has low drag.

(v) The nose-down pitch resulting from a two-point landing helps to shed lift and prevent bouncing, although lift dumping may still be required.

6.5.1.2 Tailwheel Layout

The tailwheel arrangement is the traditional one, but it is not frequently used on current designs. The main advantages are:

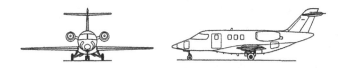

NOSEWHEEL (TRICYCLE) LAYOUT

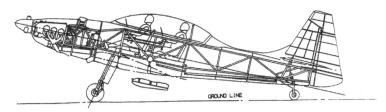

TAIL WHEEL LAYOUT

Fig. 6.14 Alternative landing-gear configurations.

(i) A third support (the tailwheel) is small and light.

(ii) During braking, the aircraft tends to pitch nose down, increasing the mainwheel reaction and reducing the possibility of skidding.

(iii) There is no requirement for a tail bumper (local fuselage reinforcement) as in the tricycle configuration, because of the tailwheel.

(iv) It is often easier to mount the main undercarriage legs onto suitable structure.

6.5.1.3 Bicycle Layout

The third layout is the so-called bicycle arrangement. This has several disadvantages, but one or two particular advantages, which have resulted in its adoption in certain designs.

The important advantages of the true bicycle layout are:

(i) The main load-carrying wheels are located roughly equidistant fore and aft of the CG, thereby leaving a substantial length of the aircraft about the CG clear of obstructions. This is particularly useful for bomber aircraft or vertical take-off aircraft where payload or engines need to be near the CG and would compete for space with the retracted position of the main gear of a tricycle gear.

(ii) The wheels are located on the centreline of the aircraft and can be stowed in the fuselage, thereby ensuring a good wing structure. Against this must be placed the disadvantages:

  (a) Outriggers are necessary. These must be able to castor and, unless care is taken in the layout, they can become large and heavy.

  (b) The aircraft landing attitude must be carefully controlled. This is essential in roll and yaw to avoid overloading the outriggers, and in pitch to prevent excessive nose-down pitching, which is prone to occur because of the great distance of the rear mainwheel aft of the CG.

  (c) Considerable elevator power is necessary to raise the nose during take-off.

6.5.2 Wheels and Tyres

The number of tyres on each unit, the tyre pressure and the number of mainwheel leg units are all very dependent upon the permissible runway loading, but can be influenced by stowage volume and location.

Runway loading is defined in terms of equivalent single-wheel loading (ESWL) on a given mainwheel unit. Approximate allowable single wheel loads equivalent to multi-tyre units are:

Number of tyres per unit	1	2	4	8
ESWL ratio	1	1.5	2.25	2.5

These ratios assume that there is no interaction between adjacent leg units.

Torenbeek [11] gives a simple method for calculating ESWL for units with varying degrees of interaction between wheels. It also allows determination of suitable runway characteristics.

Appendix A8 gives tabular data for a wide range of standard wheels and tyres.

6.5.3 Landing Gear Layout

In order to commence design of the layout of a landing gear it is necessary to be in possession of certain information:

(i) Type of layout, including number of leg units (tricycle, tailwheel, etc.).

(ii) Tyre layout on each unit and tyre pressure.

(iii) Approximate tyre size and axle travel.

(iv) Aircraft (wing) angles at landing and take-off.

(v) Extreme variations in the vertical and horizontal locations of the CG.

The most satisfactory procedure is to consider first the wheel disposition in elevation for the landing and take-off conditions, and then the arrangement in plan. A nosewheel undercarriage is considered to be the most common form.

6.5.3.1 Layout in Elevation: Landing Case

This stage of the process is illustrated in Fig. 6.15. Knowledge of the wing incidence on the approach enables the ground line to be drawn relative to the aircraft datum. There should normally be about 0.15 m (6 in) ground clearance unless a bumper is specifically incorporated. At touchdown the gear is extended. The position of the mainwheel, or centre of a multi-wheel assembly, is determined by ensuring that the contact point is at least 4° behind the perpendicular drawn from the ground line to the most adverse landing CG position. Excessive aft location of the wheel may imply an unacceptably large elevator power for lifting the nose at take-off, or large nose-down pitch on landing. The former is likely to be very critical on a tailless aircraft. Unduly far

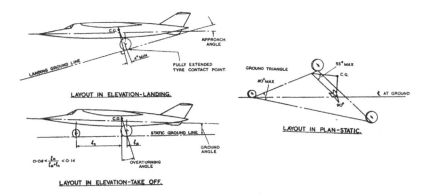

Fig. 6.15 Layout geometry limits.

forward location of the wheel implies poor ground static stability. In any case it is necessary to verify the wheel position subsequently in the take-off attitude. In a tailwheel layout the wheel contact point should be raked forward at about 17° relative to the perpendicular through the CG.

6.5.3.2 Layout in Elevation: Static Case (Take-off)

Before the static ground line can be drawn relative to the aircraft it is necessary to determine the static closure of the tyre and shock absorber. That is, the distance that the tyre and leg compress, under static loads. The former is given by the known tyre characteristics and the latter by the static spring characteristics, from comparable aircraft. The position of the mainwheels can be checked at this stage to ensure adequate ground static stability. Location of the nosewheel is initially determined by reference to the static load to which it will react. This should be about 10% of the all-up weight. Past experience has shown that less than about 8% implies poor steering adhesion and consequently difficult ground manoeuvring, whilst more than 14% results in a heavy unit and the possibility of a ground instability during taxi-ing, which is undesirable. Some variation from the initial location may be necessary to ensure adequate nosewheel mounting structure and stowage space. It may also be necessary to modify the mainwheel position.

6.5.3.3 Layout in Plan

The geometry of the undercarriage in plan is not as critical as that in elevation. It is often determined by secondary requirements such as stowage or structural attachment.

In order to avoid ground instability and nosing over during braking, the plan apex angle of the configuration should not be greater than 80°, and the angle between the ground and the line joining the CG to the side of the triangle in a plane parallel to the mainwheel track should not exceed 55°. These angles are shown in Fig. 6.15.

As a final check it is necessary to ensure that there is adequate ground clearance in all likely landing or ground attitudes. Landing with, say, 5° of bank is especially critical on aircraft with swept wings where ailerons, flaps, or underslung powerplants aggravate the situation. In all cases there should be clearances of the order of 0.15 m (6 in) for fixed parts or about twice this for moving parts.

More detailed layout information may be found in the authoritative book by Conway [12] or the more modern book by Currey [13].

6.5.4 Undercarriage Retraction

All but the simplest aircraft use retractable undercarriages to reduce the drag of the aircraft in cruise conditions.

This often leads to some of the most difficult layout problems in aircraft design. Figure 6.16 shows the complex geometry required for retraction of the gear of a low-wing aircraft, avoiding intrusion into the structural wing-box.

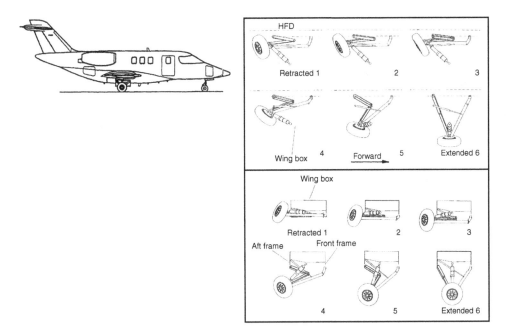

Fig. 6.16 The retraction sequence of the Cranfield E-92 project.

6.6 RECENT INNOVATIONS IN AIRFRAME SYSTEMS

Hydraulic power systems are efficient, but are significant sources of unreliability. Air bleed from engines is used in most current environmental control systems, but at significant fuel burn cost. These considerations, and others have led to the concept of "More Electric Aircraft". Secondary power is provided by more powerful electrical generators and the hydraulic system is eliminated, as is the need to bleed air from the engines. This places greater demand on electrical actuators for flying controls and landing gear actuation. One of the two main types is the electro-hydraulic actuator (EHA), which has its own internal hydraulic system. This eliminates the heavy and complex central hydraulic system, and its associated pipework. Fig. 6.17 shows a schematic of such a system.

Figure 6.18 shows the alternative electro-mechanical actuator (EMA). This is simpler, but must be provided with anti-jamming devices, for safety.

Figure 6.19 shows an image of the more electric environmental control system of the Cranfield A-9 airliner project. Cold air is taken into the cabin by ram air intakes in the nose of the aircraft. Electrical motors, under the flight deck floor, drive electrical compressors, ozone converters and air packs. The hot and cold air is mixed and filtered and then distributed by conventional distribution ducts, as shown on the upper portion of the figure. The fuselage cross-section shows the airflow patterns in the passenger and cargo compartments.

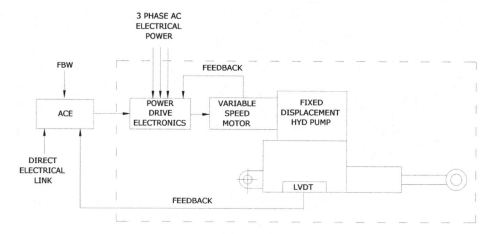

Fig. 6.17 Electro-hydraulic actuator.

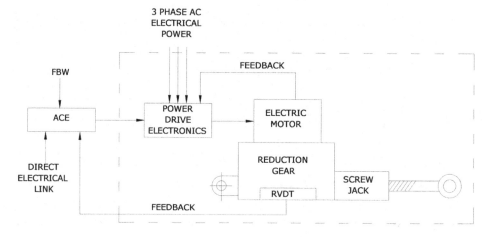

Fig. 6.18 Electro-mechanical actuator.

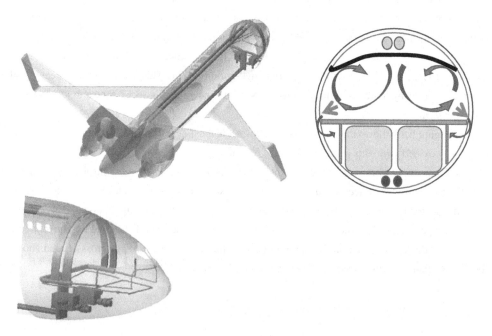

Fig. 6.19 The Cranfield A-9 electrical environmental controls.

CHAPTER 7

WHAT'S UNDER THE SKIN?: AVIONICS, FLIGHT CONTROL AND WEAPON SYSTEMS

7.1 AVIONIC SYSTEMS

Avionics is one of the most rapidly developing fields of aircraft design. Its importance and range has increased over recent years and as much as 40% of the cost of a new aircraft can be attributed to avionics. There is a bewildering range of avionic systems, each of which usually requires the use of many acronyms.

Figure 7.1 shows an avionics fit for a medium-range airliner of the future. This aircraft is intended for flight with a crew of one or two pilots. If this aircraft had been in operation during the early post-war years, three additional crew-members would have been required, namely navigator, wireless operator and flight engineer. Modern efficient, reliable avionics have dramatically simplified these operations and have eliminated to need for these crew members.

The growth in the capabilities of military avionics has been even more dramatic and make it possible for pilots of single-seat aircraft to navigate, communicate, detect and attack targets at heights of 100 ft and speeds approaching the speed of sound. Figure 7.2 shows a schematic of such an avionic system whilst Fig. 7.3 shows the installation of components on the instrument panel.

Having looked at typical aircraft installations, the next stage is to examine the functions of various types of avionic systems in the following major groups: communications, navigation systems, radar systems and others.

7.1.1 Communications

Airborne communications systems vary considerably in size, weight, range, power requirements, quality of operation and cost, depending upon the desired operation.

The most common communications system in use today is the very high frequency (VHF) system. In addition to VHF equipment, large aircraft are usually equipped with high frequency (HF) communications systems and some are fitted with ultra high frequency (UHF) systems.

VHF airborne communication sets operate in the frequency range from 100 to 150 MHz. VHF receivers are manufactured that cover only the communications frequencies, or both communications and navigation frequencies. In general, the VHF radio waves follow approximately straight lines. Theoretically, the range of contact is the distance to the horizon and this distance

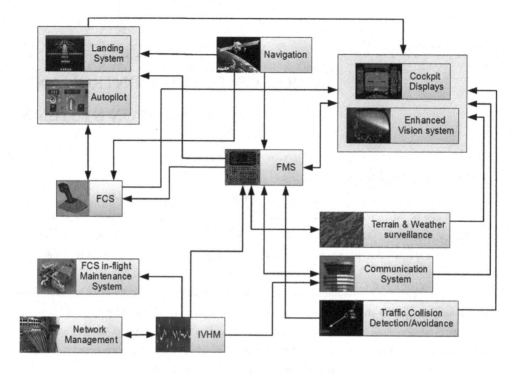

Fig. 7.1 Civil aircraft avionic systems architecture. Most of the acronyms are explained in this chapter, except for: FCS = flight control system, FMS = flight management system, IVHM = integrated vehicle health management system.

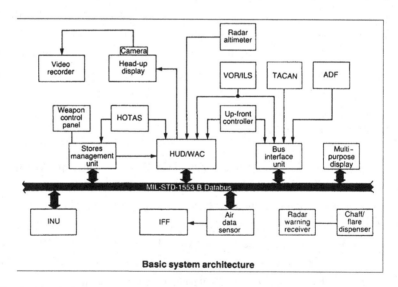

Fig. 7.2 Hawk 200 avionic system architecture. Most of the acronyms are explained in this chapter, except for: HOTAS = hands on throttle and stick, INU = inertial navigation unit, HUD/WAC = head up display/ weapon aiming computer.

Fig. 7.3 BAE SYSTEMS Hawk cockpit displays.

is determined by the heights of the transmitting and receiving antennas. However, communication is sometimes possible many hundreds of miles beyond the assumed horizon range. Typical ranges are 200 miles at 20 000 ft. UHF systems are similar to VHF but operate in the 200–400 MHz band.

An HF communication system is used for long-range communication. HF systems operate essentially the same way as VHF systems, but in the frequency range from 2 to 30 MHz. Communications over long distances are possible with HF radio because of ionospheric reflection. HF transmitters have higher power outputs than VHF transmitters.

The design of antennas used with HF communication systems vary with the size and shape of the aircraft. Aircraft that cruise below 300 mph generally use a long wire antenna between the fuselage and probes installed on the fin.

Air traffic controllers need to know the positions of all aircraft within their control areas. Ground-based radar systems detect all aircraft, but some system is required to identify them. Civil aircraft carry equipment called an ATC transponder, which receives pulses from the ground radar sets. The transponder then transmits a sequence of pulses, which identifies the aircraft and gives its height to the controller.

This system is a development of the military identification friend or foe (IFF) system. The IFF responds to signals from either ground-based or airborne radars. This has an obvious military value.

7.1.2 Navigation Systems

The knowledge of aircraft position has always been important in flying and initially involved visual 'fixes' made in good visibility, with the aid of a compass. This method is still used by some light aircraft pilots, but increasingly sophisticated navigation systems have been developed.

7.1.2.1 Automatic Direction Finding (ADF)

The ADF principles may be illustrated by listening to a transistor radio. The aerial is directional and the signal becomes weaker or stronger as the radio is rotated about a vertical axis. On aircraft, loop aerials are rotated in the direction of ground-based non-directional beacons (NDBs). An aircraft is usually fitted with two systems, each of which automatically detects the position of two NDBs. The signals from each system are shown to pilots on two radio magnetic indicators (RMIs). The RMIs give the magnetic bearing of the NDBs, which will give information to locate the aircraft on a map. Maximum range is 10–150 nautical miles (n.miles) with a typical accuracy of $\pm 4°$.

7.1.2.2 VHF Omnidirectional Range (VOR) System

The VOR is an electronic navigation system. As the name implies, the omnidirectional or all-directional range station provides the pilot with courses from any point within its service range. It produces 360 usable radials or courses, any one of which is a radio path connected to the station. The radials can be considered as lines that extend from the transmitter antenna like spokes of a wheel. Operation is in the VHF portion of the radio spectrum (frequency range, 108.0–117.95 MHz), with the result that the radio interference from atmospheric and precipitation static are negligible. The navigational information is visually displayed on an instrument in the cockpit. The range is up to 130 n.miles at 12 000 ft with an accuracy of $\pm 1°$.

7.1.2.3 Distance Measuring Equipment (DME)

This is an interrogation and response system, where the aircraft equipment sends a VHF signal to a ground station, which then responds to the aircraft. The DME receiver computes the time delay and calculates the distance from the beacon. This system is often combined with VOR, which supplies the direction from the beacon, and DME gives the distance. The range is similar to that of the VOR, with an accuracy of ± 0.5 nmiles.

The above system is used in civil operations, but its military equivalent is TACAN (tactical air navigation). This combines features of both VOR and DME, and gives range and bearing from a beacon.

7.1.2.4 Doppler Navigation

This is an obsolescent medium- to long-range self-contained system. A typical system accuracy might be 1% in distance and heading, which would be serious at long ranges unless corrected by other systems.

The Doppler radar emits narrow beams of energy at one frequency, and these waves of energy strike the Earth's surface and are reflected. Energy waves returning from the Earth are spaced differently than the waves striking the Earth. The Earth-returned energy is intercepted by the receiver and compared with the outgoing transmitter energy. The difference, due to Doppler effect, is used to develop ground speed and wind drift angle information.

7.1.2.5 Long Range Navigation (LORAN) and Omega

LORAN has been a major navigation system that utilizes the hyperbolic principle. Several hyperbolas are produced by means of 100 kHz pulses transmitted from the master ground stations. Position is fixed by using several ground stations to produce hyperbolic systems whose intersections describe unique points, as shown in Fig. 7.4.

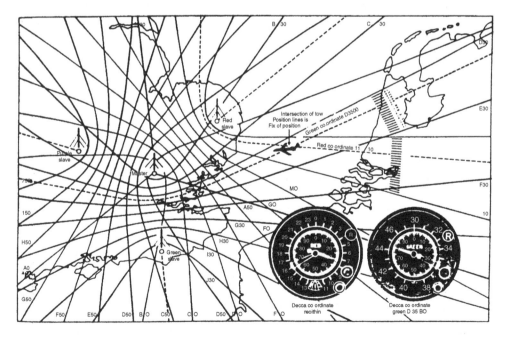

Fig. 7.4 LORAN navigation hyperbolae lines.

At close range to the station the 'groundwave' is predominant (the directly transmitted signal) and has a range of 600 nmile. At longer ranges the 'skywave' (the signal reflected from the ionosphere) is used and ranges of 2500 nmile can be achieved. A more modern system is Omega, which uses eight ground stations world-wide and uses VLF transmitters.

7.1.2.6 Global Positioning by Satellite (GPS)

This is a relatively recent system, which is being increasingly used by civil and military aircraft. It takes position fixes from signals from a global array of geo-stationary satellites. It is phenomenally accurate, lightweight and relatively cheap. Accuracies to tens of metres may be achieved. It is displacing some of the older navigation systems and is being considered for use in instrument landing systems.

7.1.2.7 Inertial Navigation

A typical inertial navigation system (INS) measures inertial forces, or the effects of accelerations, with accelerometers mounted on a stabilized platform. Basic stabilization is achieved gyroscopically, but corrections have to be made for the effects of the Earth's rotation, and Coriolis and centrifugal forces.

Accelerations are resolved into north–south and east–west components (and sometimes also vertically), and then integrated successively to provide, at any time during the flight, instantaneous speed and accumulated distance from take-off. Given the geographical coordinates of its starting point, an INS can provide a position readout at any time during flight. Accuracy deteriorates with time, but in a good system, the error accumulates at less than 1 nmile/h. Speed and distance can be processed to provide true airspeed, wind velocity, track, heading, cross-track error and distance to go. INS can be coupled to the autopilot (see below) to provide automatic track-keeping.

For fail-safe operation up to three independent INS sets may be fitted, with the crew comparing the output of individual sets to obtain an accurate position.

7.1.2.8 Area Navigation

In an area navigation (RNav) system, bearing and distance, position or velocity information from such aids as VOR is processed in a computer. This produces steering information between waypoints, which are programmed into an autopilot system, prior to take-off (way-points are geographical positions where heading changes are made).

A large range of equipment is available. Simple units – usually produced for general aviation applications – use two or more receiver frequencies simultaneously (any combination of VORs or DMEs) and a computer. The most sophisticated RNavs embody larger computers, usually with a large back-up memory capable of storing details of the operator's route structure. Autopilot coupling and flight-director outputs are common in such systems, and there may be electronic data displays and a moving-map image.

7.1.3 Radar Systems

Radar is a device used to see certain objects in darkness, fog or storms, as well as in clear weather. In addition to the appearance of these objects on the radar scope, their range and relative position are also indicated.

Radar is an electronic system using a pulse transmission of radio energy to receive a reflected signal from a target. The received signal is known as an echo; the time between the transmitted pulse and received echo is computed electronically and is displayed on the radar scope in terms of range in nautical miles.

A radar system consists of a transceiver and synchronizer, an antenna installed in the nose of the aircraft, a control unit installed in the cockpit and an indicator or scope. A wave-guide connects the transmitter/receiver to the antenna.

7.1.3.1 Weather Radar
In the operation of a typical weather radar system, the transmitter feeds short pulses of radio-frequency energy through a waveguide to the dish antenna in the nose of the aircraft. Part of the transmitted energy is reflected from objects in the path of the beam and is received by the antenna. Most weather radars have alternative modes to provide simple ground-mapping information.

7.1.3.2 Attack Radar
Attack radars operate in a similar manner to the weather radar described above. The main difference is that attack systems use more power and are used to detect small, fast-moving targets rather than weather formations. Most systems are also used to guide semi-active radar homing missiles and must be designed to be resistant to enemy jamming.

7.1.3.3 Terrain-Following Radars
Military ground-based radars give good information about attacking aircraft. The curvature of the Earth and hills, however, leave a 'blind-spot' at very low altitude. Attack aircraft therefore exploit this loophole and attack at extremely low level at high speed. This obviously increases the danger of inadvertently flying into the ground. Therefore an independent high-power system is used that detects ground obstacles and feeds control signals into the aircraft's flight control system. It requires only a narrow beam width.

7.1.3.4 Airborne Early-Warning Radars
This is a system designed to overcome the low-level blind-spot of ground-based radars. In this system, powerful radar sets are flown in long-endurance aircraft. The altitude of the aircraft is such that the range of the radar is much greater than with ground-based systems and even low-flying aircraft can be detected at greater distances.

7.1.4 Other Avionic Systems

Many other avionic systems are used by both civil and military aircraft. The major ones will not be fully described here, but merely listed for completeness.

7.1.4.1 Autopilot

This is a system of automatic controls that return an aircraft to a pre-set flight pattern when it is displaced from it. The autopilot interfaces directly into a powered flight control system, to control the flying surface actuators. Autopilots for aircraft with manual flight controls have their own actuators which supply inputs into the control circuits.

7.1.4.2 Instrument Landing Systems

The instrument landing system (ILS), one of the facilities of well-equipped airfields, operates in the VHF portion of the electromagnetic spectrum. It can be visualized as a funnel made of radio signals on which the aircraft can be brought safely to the runway.

   The entire system consists of a runway localizer, a glide slope signal, and marker beacons for position location. The localizer equipment produces a radio course aligned with the centre of an airport runway. The glidescope indicates the correct vertical approach path for the aircraft, which is usually a 3° slope for civil aircraft. An extension to this system is autoland, in which the aircraft is automatically landed without any pilot input. More recent developments have led to the more accurate microwave landing system (MLS).

7.1.4.3 Laser Rangers and Target Markers

Laser beams are used for weapon aiming in which a target is marked by a laser. A suitable missile or bomb then flies along the beam to the target.

7.1.4.4 Forward-Looking Infra-Red (FLIR)

This is a system which detects heat produced by a target. This might be the engines or a relatively hot part of the aircraft structure. An infra-red-seeking missile can be directed onto the target. These systems are not, strictly speaking, avionic systems, but signals from them are fed into weapon-aiming systems, which *are* avionic systems.

7.1.4.5 Low-Light Television (LLTV)

This is an optical system in which targets can be seen in clear air at night. This is done by using image intensifiers attached to a TV camera to produce a day-bright picture of the ground.

7.1.4.6 Electronic Counter-Measure (ECM)

These are systems that either generate fake signals to confuse enemy radar systems or produce a high-powered blanket signal to jam enemy reception.

7.2 FLIGHT CONTROL SYSTEMS

The control system consists of four basic elements, namely the pilot or autopilot, the linkage, actuators and the control device itself. The autopilot in an aircraft or missile computes the motion of the vehicle with the aid of accelerometers and gyros, and stabilizes it on a given flight path until it is over-ridden by a guidance command. It is essentially a device for measuring deviations from a desired flight path that possesses the means for providing a correction. In a missile it is often convenient to combine it with the guidance electronics package. The control actuators convert the signal from the autopilot into a motive force that is used to actuate the control device.

The control system must provide sufficient force for the vehicle to manoeuvre in accordance with reasonable guidance commands, but it must be able to do this in a sufficiently short time to enable these demands to be kept to a reasonable level. The value of g that a vehicle is required to achieve will depend upon its type and the form of guidance employed. For example, a large aircraft may have a flight g capability of 2.5, and interceptor aircraft may be able to develop 9 g, and an anti-aircraft missile 15 g. This manoeuvrability is a function of wing lift and above a certain altitude it decreases with increasing height. The rate at which the control system can cause the vehicle to achieve the required g is a function of its aerodynamics, structural design and control device.

7.2.1 Conventional Trailing-Edge Flying Control Surfaces

Figure 7.5 gives an illustration of the position of flying controls on an aircraft. They are categorized into three groups.

7.2.1.1 Primary Group
The primary group includes the ailerons, elevators and rudder. These surfaces are used for moving the aircraft about its three axes.

The ailerons and elevators are generally operated from the cockpit by a control stick on a single-engine aircraft, and by a wheel and yoke assembly on a multi-engine aircraft. The rudder is operated by foot pedals on all types of aircraft.

7.2.1.2 Secondary Group
Included in the secondary group are the trim and spring tabs. Trim tabs are small aerofoils recessed into the trailing edges of the primary control surfaces. The purpose of trim tabs is to enable the pilot to trim-out any unbalanced condition that may exist during flight, without exerting any pressure on the primary control surface. It is operated by an independent control, which is a useful back-up system, in the event of primary system failure.

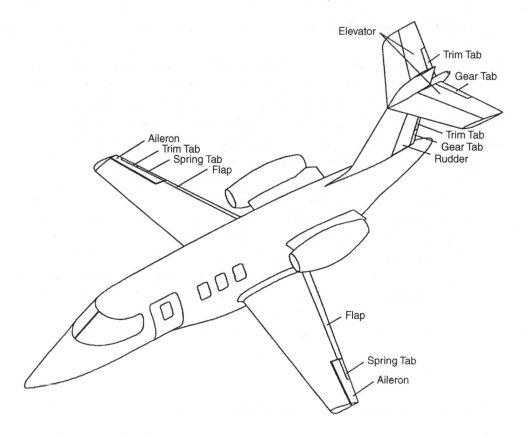

Fig. 7.5 Flight control surfaces.

Spring tabs are similar in appearance to trim tabs, but serve an entirely different purpose. Spring tabs are used to reduce pilot effort in moving the primary control surfaces. Tabs are not used on larger, high-performance aircraft with powered controls.

7.2.1.3 Auxiliary Group

Included in the auxiliary group of flight control surfaces are the wing flaps, spoilers, speed brakes and slats.

Wing flaps are used to give the aircraft extra lift and are hinged to either the trailing edge or the leading edge of the wing. Wing flaps on most older aircraft were hinged to the trailing edge of the wing inboard of the ailerons. Some later aircraft have flaps hinged to both the trailing edge and the leading edge of the wings and may be deployed on tracks to increase the chord. Their purpose is to reduce the landing speed, thereby shortening the length of the landing run, and to permit landing in small or obstructed areas. In addition, the wing flaps are used during take-off to shorten the take-off run, or in combat to improve manoeuvrability.

Spoilers are air brakes for the purpose of decreasing wing lift; however, their specific design, function and use vary on different aircraft. On some aircraft the spoilers are long narrow surfaces hinged at their leading edge to the upper surface of the wings. In the retracted position, the spoiler is flush with the wing skin. In the raised position, wing lift is greatly reduced by destroying the smooth flow of air over the wing surface.

Spoilers are sometimes used differentially, to assist or replace ailerons as roll control devices.

7.2.2 All-Moving Tailplanes or Foreplanes

Some aircraft use a movable horizontal surface called an all-moving tail. This serves the same purpose as the tailplane and elevator combined. When the cockpit control is moved, the complete surface is moved to raise or lower the leading edge, thus changing the angle of attack and the amount of lift on the tail surface.

As this is a larger control surface than the elevator, it has a more powerful effect.

The foreplane of an aircraft with what is called a canard configuration uses a pitch-control device *ahead* of the wing, rather than behind it. This has the advantage that in most cases the trimming lift is upwards, thus assisting the wing in producing lift. Conventional tailplanes often produce downwards-acting lift to trim-out pitching moments. Canard arrangements, however, generally exhibit more stability and control problems than do conventional layouts.

It is possible to activate all-moving tailplanes differentially to act as roll control devices. These are then termed 'tailerons'.

7.2.3 Reaction Devices

There are various ways of producing control forces by reaction means. For example, moving fins may be in a rocket exhaust or the nozzle may be moved in some way, so that changes in thrust direction are obtained. Liquid rockets often have tiltable combustion chambers, with small variable thrust units as a vernier control. Solid rockets may have rotating, canted, nozzles that are used in groups of three or more. Tilting a nozzle, as distinct from a combustion chamber is also a possibility, but does introduce mechanical complications. Reaction controls are essential on very high-flying vehicles where the air density is too low to enable sufficient aerodynamic force to be developed. Reaction controls are also used in vertical take-off and landing aircraft such as the Harrier. In this case the aircraft has insufficient forward speed to make conventional controls effective.

7.2.4 Linkage between Pilot and Control Surfaces

The flying control linkage once consisted of a simple arrangement of levers and wires (Fig. 7.6). The situation in a modern high-speed aircraft is very different. Power controls are necessary, and the control system is often reduced to a signalling device with artificial feedback on control forces to give the pilot some idea of the loads being imposed on the aircraft. The situation is complicated

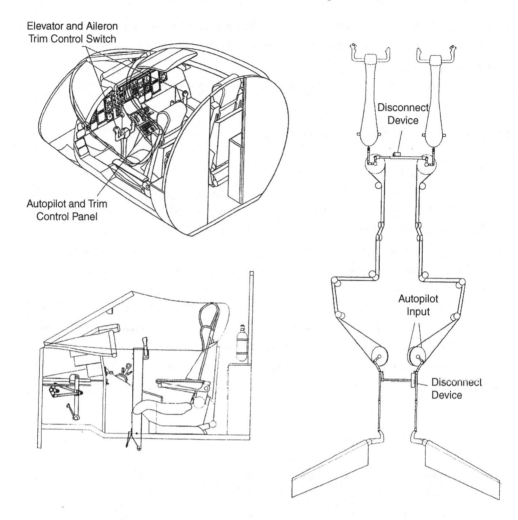

Elevator and Aileron
Trim Control Switch

Disconnect
Device

Autopilot and Trim
Control Panel

Autopilot
Input

Disconnect
Device

Fig. 7.6 Control system links from flight deck to the elevators.

by the need for autopilots, artificial stabilization, automatic landing, etc. The tendency is for the control system to become fully automatic, with the pilot simply being present to monitor the system and provide an element of initiative in the case of a fault. In this respect the control is comparable with that of a guided missile. Since a control failure cannot be tolerated, it is necessary to resort to a multiplexed system. There appear to be many advantages in using fully electrical signalling all the way to the actual control actuator, as in the F-16 fighter.

Advances in computational power have led to the development of 'active' controls, which are currently used in the following applications.

7.2.4.1 Relaxed Static Stability (RSS)
Normal aircraft have a tail device more than large enough to provide natural longitudinal static and dynamic stability. This is in some conflict with crisp manoeuvrability. The more stable an aircraft,

the less manoeuvrable it is. RSS is used to reduce the tailplane size below that required for natural stability. This has the effect of increasing manoeuvrability, and reducing the mass and drag of the tailplane or foreplane. Stability, however, is significantly reduced and may indeed be negative. The flight control system, therefore, has to sense any impending destabilizing event (say a gust) and then rapidly send a correcting signal to the elevator, tailplane or foreplane. It is usual to have three or four independent electrical signalling systems. Fibre-optical signalling is being developed as an alternative.

7.2.4.2 Manoeuvre Load Control (MLC)

The spanwise airload is modified by use of the ailerons and, possibly, flaps to reduce wing bending moment and shear, as shown in Section 5.2.3, above. Manoeuvres are relatively slow-acting phenomena, so rapid response is not required, unless the system is combined with gust load alleviation, below.

7.2.4.3 Gust Load Alleviation (GLA)

This is the rapid reconfiguration of the wing to alleviate the effects of gusts. This is also described in Section 5.2.3.

7.3 Weapon Systems

7.3.1 Guns

Aircraft cannon are effective in attack against all types of vehicles, field guns, aircraft on the ground, moving troops and ship superstructures. They are, however, limited in range to about 800–1000 m air-to-air and 2000 m air-to-ground.

Smaller calibre machine guns have even less range and their use is limited to armed light aircraft and helicopters. There are two categories of aircraft cannon.

7.3.1.1 Gas-Operated Revolver

These guns use a single barrel combined with a four or five chamber rotary revolver, very much like the cowboy's six-shooter. The operation of the gun is powered by the gas discharged during the firing of the first round.

When the first round is fired, by means of an electrical signal, the exhaust gas revolves the breech and removes the used case and link. Guns which use this principle include the ADEN, DEFA, OERLIKON and guns from the former Soviet Union. Figure 7.7 shows the former and their performance is shown in Table 7.1. Their rates of fire are considerably less than for the rotary guns, but they tend to be lighter and less bulky.

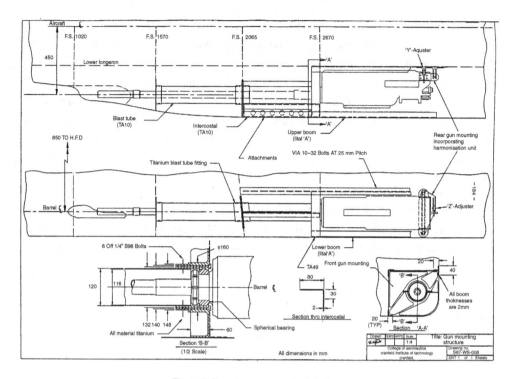

Fig. 7.7 Revolver cannon installation.

7.3.1.2 Rotary 'Gatling' type

These guns range in calibre from the 7.62 mm mini-gun through the popular 20 mm Vulcan cannon to the powerful 30 mm GAU–8 (Fig. 7.8) fitted to the Fairchild A-10 aircraft shown in Fig. 4.9. This type of gun has a varying number of identical barrels ranging between three and seven. External power is required to rotate the barrel cluster, and rotate a cam to feed and fire the ammunition. External power may be electrical, hydraulic or pneumatic. The ammunition is, again, either electrically or percussion operated. The former method may be sensitive to electromagnetic interference.

7.3.2 Bombs

These are the primary weapons for stationary targets, for the destruction of runways, oil dumps and other installations. Dimensional information is given in Appendix A.

7.3.2.1 Conventional Free-Fall Bombs

The classic free-fall bomb has been used from aircraft since World War I and is now usually in the 1000 lb category. The dimensions of this weapon are length 90" (2.26 m) and diameter 16.5" (0.42 m).

Table 7.1   Gun performance

Gun type	ADEN/ DEFA	Oerlikon 304RK	Vulcan M61	GE–525	GAU–8	GE–430
Calibre (mm)	30	30	20	25	30	30
Gun length (m)	1.66	2.69	1.88	2.11	2.88	2.79
Muzzle velocity (m/s)	815	1075	1036	1100	1040	1040
Rate of fire (rounds/s)	21	23	100	60	70	40
Shell weight (kg)	0.236	0.325	0.1	0.18	0.37	0.37
Round weight (kg)	0.363	0.664	0.254		0.70	0.70
Shell weight/s	4.96	7.48	10.0	10.98	25.9	14.8
Installation weights – gun + 5 s mass of ammunition fired (kg)						
Gun	80	125	115	122	282	136
Feed Mechanism					250	
Transfer Unit					20	
Drive System					30	
Ammunition				152	243	139
Total weight (kg)					825	
(Gun + 5 s)						
Recoil force (N)		20 000	17 200			

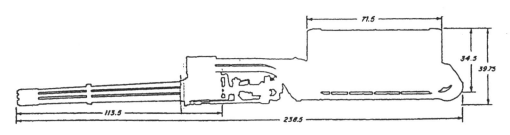

Fig. 7.8 GAU–8 cannon.

Whilst being a reasonable aerodynamic shape, they also produce a considerable amount of drag. A range of low-drag conventional bombs has been developed. They are more slender than conventional types, but are necessarily longer.

The fusing arrangements within the case are designed to ensure that the bomb is inert during loading and carriage. A mechanical safety-pin prevents the fuse from functioning during the transit flight. After the bomb is released from the aircraft, the pin is withdrawn by a lanyard and a timing mechanism is set in operation to arm the bomb.

Most free-fall bombs use a high explosive charge, but it is possible to use a nuclear or chemical warhead.

7.3.2.2 Retarded Bombs

In low-level free-fall bombing there are two problems: first there is the danger that the aircraft may itself be damaged by the bomb explosion; second, the trajectory of the bomb is very flat and the bomb tends to bounce or 'skip' after impact. These two difficulties have been overcome by fitting the bomb with a retarder tail. This is a drag-increasing device opening somewhat like an umbrella which slows the bomb down and steepens its trajectory so that it impacts the ground some distance behind the aircraft. This does not completely eliminate skipping, and therefore instantaneous fusing is used to explode the bomb on initial impact. To protect the aircraft, bomb deceleration is measured and a sensor prevents the fuse from being activated, if the retarder fails to open. Retarded bombs are normally ejected from altitudes of between 200 and 500 ft.

7.3.2.3 Cluster Bombs

The cluster bomb is a versatile weapon that can cope with a variety of targets. It comprises a large number of small bomblets enclosed in a bomb-shaped casing. Each bomblet has its own sensor and fusing device. After the bomb is released, a charge explodes the casing and a further charge ejects the bomblets in a controlled pattern so that jostling between them is avoided. The cluster bomb is effective against both hard and soft targets, but is now banned from use by the International community.

7.3.2.4 Laser-Guided Bombs

Laser target marking and laser spot-seeking bombs have been in use for more than 30 years. In this system, special 'smart' bombs are fitted with a sensor in their nose, which seeks the laser spot trained onto the target by a ground or airborne target designator and homes the bomb onto the spot. The bomb is given a limited degree of manoeuvrability in both azimuth and elevation by control surfaces replacing the fixed fins.

7.3.3 Rockets

Rockets provide greater hitting power than guns and when used in a dive attack, are highly lethal and economic relative to guided missiles. However, the aircraft diving from altitude is vulnerable

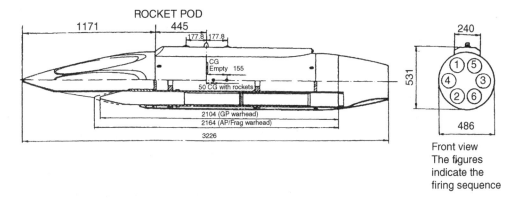

Fig. 7.9 Underwing rocket pod. (Dimensions in mm.)

to attack from ground defences. In a low-level attack, rockets are less effective because of the difficulties of gravity drop and of precise aiming.

A typical modern six-round pod is shown in Fig. 7.9. This uses large unguided 135 mm rockets with a typical range of 3 km. Smaller 68 mm rockets are available in pods containing 18 or 19 rounds.

These unguided rockets have a shorter range and must be used in situations where there is little opposition, or heavy aircraft losses will follow. They are more useful against a large spread-out target. A typical aircraft installation would include four pods.

7.3.4 Guided Missiles

7.3.4.1 Missile Guidance: Surface-to-Surface and Air-to-Surface

Although a missile is unlike a manned aircraft in that it is designed for only one flight, the most important and fundamental difference is in its guidance system. Even in a modern, sophisticated aircraft with many automatic flying aids there is always a pilot to monitor the behaviour of the machine. A successful missile must be capable of destroying its target without this human monitoring, at least as far as the actual flying vehicle is concerned.

There are numerous types of guidance, each of which is suited to a particular application. When attacking a static target the missile requires only to know its position relative to the target, or to some known origin which is normally derived from a flight programme fed in before launch.

One of the simplest types of guidance systems is that of using a predetermined magnetic heading. An autopilot in the missile receives information from a magnetic sensing element and keeps the missile on the desired course.

Other guidance systems used are similar to those of manned aircraft, i.e. radio navigation and inertial navigation. The latter may be updated by star tracking.

Terrain comparison (TERCOM) is a recent development in which sensitive altimeters measure the profile of the ground beneath it and check it against pre-programmed information.

This system is combined with inertial systems to give extreme accuracy and is used on cruise missiles. The system, however, requires good intelligence about potential targets and is therefore inflexible.

7.3.4.2 Missile Guidance: Surface-to-Air and Air-to-Air

(i) Command Systems

In a simple command system, the missile is controlled by an operator who is located in a position suitable for sighting the target. This position need not necessarily be at the launching site. The operator observes both the target and missile, and guides the missile accordingly. Flares are often placed onto the missile to assist the visual tracking. The control button is often incorporated in a monocular or binocular sight.

There are three usual types of link between the operator and the missile, wire, radio and television,

(ii) Command Link System

A command link system is a more elaborate development of the simple command system and aims at removing its limitations. The target is tracked by ground-based radar and its velocity and position are computed. A similar installation tracks the missile and computes its motion. The results for the target and missile are then compared automatically and a radio transmitter sends control signals to the missile to enable it to engage the target. The principle is illustrated in Fig. 7.10.

(iii) Beam-Riding

A beam is pointed towards the target and the missile is controlled to fly along this, either a surface or an airborne installation being used to direct the beam.

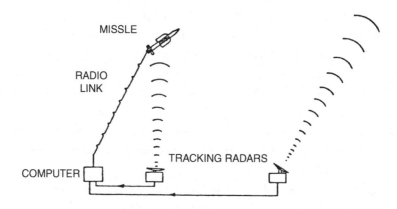

Fig. 7.10 Command link guidance.

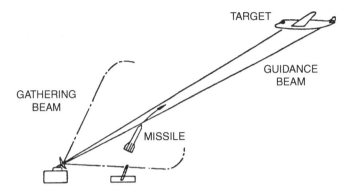

Fig. 7.11 Beam-riding guidance.

It is normal to use a radio beam, but, with the development of lasers, which are capable of producing intense, narrow, light beams, these give a good alternative (see below).

There is a range limitation on accuracy, as the beam width increases in direct proportion to the distance from the source. It is often not possible to launch the missile directly into the main guiding beam and when this is so a wide-angle 'gathering' beam is used initially, as shown in Fig. 7.11. The missile guidance components are relatively simple and any number of missiles can be directed along one beam. Another advantage is that surface installations can be readily moved.

(iv) Homing Guidance

In a homing guidance system the missile itself has a device for looking at the target. With this, the accuracy of the system will tend to increase as the target is approached.

The size of the reflector in the homing head may be decided either by the range or by the discrimination necessary against multiple targets, which may determine beam width at a given range. There are three variations of a homing guidance system, shown in Fig. 7.12.

Active homing is the most complex (and expensive!). The missile is equipped to transmit radar signals in the direction of the target. The reflected signals obtained enable the target to be followed by the homing head, which computes the required control signal. This missile is independent of external guidance.

The semi-active radar target is illuminated by a transmitter located away from the missile. The missile receives the reflected signals as in the case of active homing. Guidance in the missile is less complex as it carries only the receiving apparatus. A number of targets being attacked by a number of missiles can be illuminated by one transmitter, and this transmitter can be mobile. Range can be relatively large, since the transmitter radar can be relatively large.

Semi-active laser guidance uses the same principle, but the target is illuminated by a laser, rather than radar. The system may be used for missiles or 'smart' bombs (above).

The most common forms of passive homing systems employ the use of infra-red heat-seeking cells. Alternatively, they may home on the acoustic properties of the target, in a similar

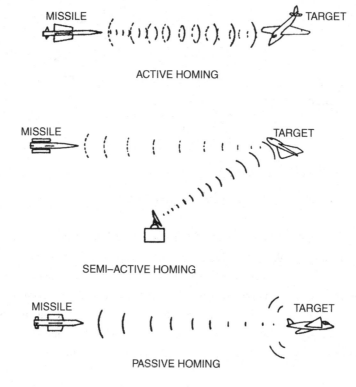

Fig. 7.12 Homing guidance systems.

fashion to torpedoes. The absence of the need for a transmitter is the main advantage and the result is a flexible system, which like the active homing system has training advantages. Enemy counter measures are usually limited to providing an alternative source of energy emission. Infra-red systems are suitable for day and night operation, but the presence of clouds at lower altitudes may prove to be a difficulty in some cases. Homing is usually on the exhaust from aircraft engines, but recent developments of more sensitive units may enable other sources of lower temperature to be used.

Both active and passive guidance systems are termed 'fire-and-forget' systems. The missile operators need to obtain either radar or infra-red lock-on prior to launch. The missiles are then guided onto the target automatically, leaving the operator to turn his attention to alternative targets.

7.3.4.3 Typical Air-to-Ground and Air-to-Air Missiles

The Maverick is a relatively large anti-tank weapon that is aimed by television, semi-active laser or semi-active infra-red. The television-guided version requires a second crew member to guide it, but the other versions are of the 'fire-and-forget' type suitable for launching from single seat aircraft. This does, however, imply a comprehensive avionics fit. It would be possible to fire four semi-

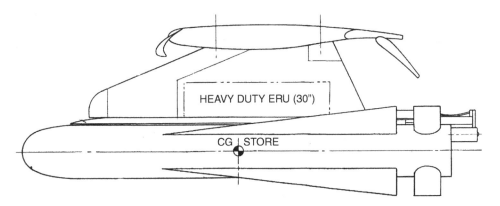

Fig. 7.13 Maverick missile.

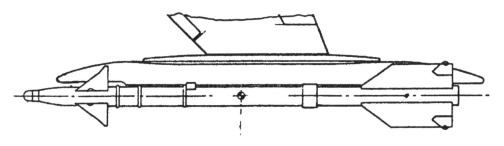

Fig. 7.14 Sidewinder missile.

active versions on a single attacking manoeuvre. This weapon has a range of greater than 7 km (Fig. 7.13).

The Sidewinder is a widely used short-range weapon with a range in excess of 5 km. Figure 7.14 shows the missile, complete with its launcher. These missiles are often placed on aircraft wing tips, where their end-plate effect reduces induced drag, but the wing has to be sufficiently stiff to carry them. The missile is suspended from three points.

CHAPTER 8   WHY DO AIRCRAFT COST SO MUCH?

8.1 GENERAL

This chapter discusses the costs of buying and operating civil and military aircraft. The former costs are termed acquisition costs for both classes of aircraft, whilst the latter are termed operating costs for civil aircraft, and life-cycle costs (LCCs) for military aircraft. The acquisition costs are included as important elements of both operating costs and LCCs. The costs associated with aircraft reliability and maintenance are significant contributors to operating and LCCs, and design to reduce these costs is outlined at the end of this chapter.

8.2 ACQUISITION COSTS (THE COSTS OF BUYING OR ACQUIRING THE AIRCRAFT)

8.2.1 The Reasons for High Aircraft Acquisition Costs

The main reasons are:

    (i) High performance requirements – Civil and military aircraft operations are competitive, hence each new type is required to show improvements relative to existing aircraft. It is therefore necessary to undertake extensive research programmes and also to incorporate customer/operator refinements, which imply complexity.

    (ii) Safety considerations – Civil and military aircraft have, rightly, very stringent safety requirements that must be proved before the aircraft enters service. Extensive proving and testing is required and there is a need for added complexity so that failure of individual components or systems can be tolerated. An aircraft has to continue functioning safely even after a failure has occurred. This applies to both hardware and software.

    (iii) Quality control – The complexity and safety demands make it essential to have elaborate systems that record the complete history of not only each individual aircraft, but of each piece of material or part used in it. Inspection has to be complete, not sampled, and continuous throughout the life of the aircraft.

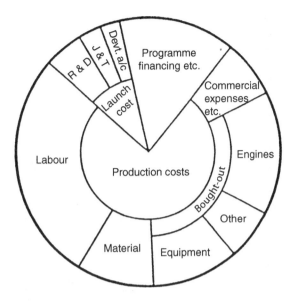

Fig. 8.1 A300 acquisition cost proportions.

(iv) Tooling requirements – Complexity requires extensive and expensive tooling for production purposes, and the launch cost of a new type of aircraft is considerably influenced by this. Numerically controlled machine tools have considerable flexibility, and are very expensive. They can more than pay for themselves, however, on reasonable production runs, and their use is increasing. Automatic rivetting and bonding facilities also reduce labour content (see below). Another source of expense is that of assembly jigs.

(v) Labour-intensive production – In spite of the elaborate tooling, the complexity of an aircraft means that labour costs are by far the greatest part of the direct production costs (Fig. 8.1). Two approaches to reducing such costs are increasing automation or use of lower-cost labour.

(vi) High overheads – Overheads on both production and operation are high because of the necessary research, development, recording, tooling, inspection and other considerations mentioned above.

(vii) Large financial investment, interest cost – Interest payments on the huge financial investment needed to launch a new design add to the cost substantially. This is aggravated by expenditure of capital before any sales income is received and the long gestation periods. Fig. 8.2 shows typical combat aircraft development times.

(viii) Short production runs – In spite of all the investment needed, the production runs are short by most industrial standards. A civil aircraft will usually be costed so that the project costs break even at production runs between 200 and 300 aircraft. This is frequently extended by the need to develop and modify the aircraft later and very few

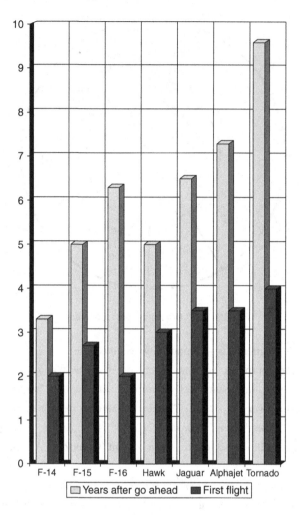

Fig. 8.2 Combat aircraft development times.

military or civil aircraft reach production of over 1000. Thus all the initial investment has to be taken by a small number of products.

The advent of integrated computer-aided design and manufacture systems (CAD/CAM) has gone some way to reducing the above problems, particularly in reducing development times and improving initial build quality.

8.2.2 Acquisition Cost Prediction Methods

The two main approaches to cost prediction are termed 'top-down' and 'bottom-up'. The former is used during the conceptual or preliminary design stages, whilst the latter is utilized during the

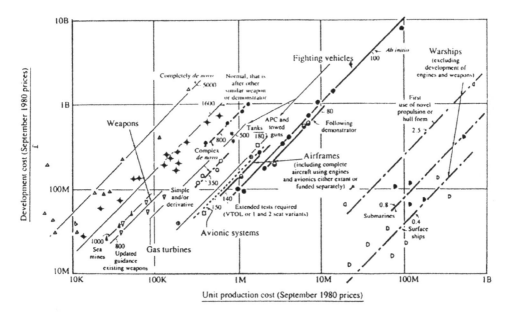

Fig. 8.3 Top-down cost estimation trends.

detail design and development phases. There is little published information about the bottom-up approach as it uses detailed analysis of every aircraft component, its material costs, production costs and overheads. These calculations are then synthesized to produce component, subassembly, assembly and whole-aircraft costs. These methods are often project-specific or contain information that is company-confidential. Jang [14] used this approach to perform a 'bottom-up' synthesis of the costs of a turbo-prop training aircraft.

There are several 'top-down' approaches, but all use statistical data from as wide a range as possible of existing aircraft. Some, such as Pugh, [15], use data for complete aircraft, as shown in Fig. 8.3. Other approaches break the aircraft costs into the major elements of airframe, propulsion and avionics, and develop prediction relationships for these. Much of the published work has its origins in work by the Rand Corporation. One example is that for the prediction of airframe costs [16].

Statistical data are gathered and analyzed to predict costs for the various elements shown below:

(i) Development costs
   Engineering.
   Tooling.
   Non-recurring manufacturing labour.
   Recurring manufacturing labour.
   Non-recurring manufacturing materials.
   Recurring manufacturing materials.
   Prototype aircraft.

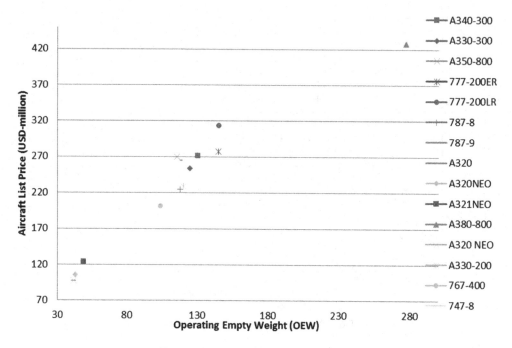

Fig. 8.4 Commercial aircraft prices, 2016.

Test specimens.

Flight test.

Quality control.

Bought-out items (from suppliers).

Financing costs.

These are incurred before the manufacture of production aircraft and must be recovered as a levy on each of the production aircraft sold, or financed by some other method.

(ii) Production costs

The elements of these costs are similar to those for development, but do not extend to prototypes or test specimens, as these are not required for production. The flight testing is much reduced, and is restricted to production and acceptance flight tests for each individual aircraft.

8.2.3 Typical Acquisition Costs

Figure 8.4 shows civil aircraft costs, based on published data from the Boeing Commercial aircraft Company, and Airbus, Ltd.

Figure 8.5 shows calculated and quoted prices of combat aircraft, based on several magazine sources, particularly *Aviation Week* and *Space Technology*.

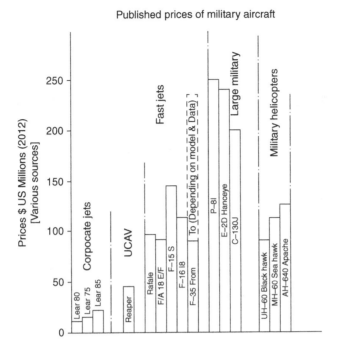

Fig. 8.5 Military aircraft prices, 2012.

8.3 CIVIL AIRCRAFT OPERATING COSTS

8.3.1 Indirect Operating Costs

Transport aircraft costs are subdivided into what are termed indirect and direct costs. The indirect operating costs (IOCs) are those associated with running an airline, rather than those directly concerned with operating aircraft. Indirect costs are the sole responsibility of an operator and account for nearly half the total costs of a scheduled operation, but are a somewhat lower portion of chartered operation costs. IOCs cover such things as central administration, sales and publicity, airport costs and passenger service. These costs, and the proportions of costs in different categories are different for each airline, reflecting the operating methods of the individual companies. Indirect costs are not normally influenced by the aircraft, but there are exceptions, for example, the cost of providing accommodation for passengers delayed by an aircraft failure would be taken as indirect.

8.3.2 Direct Operating Costs

Direct operating costs (DOCs) do vary with aircraft type and trip length. Table 3.2, in Chapter 3, shows the DOCs for a wide range of aircraft types. Figure 8.6 shows a description of all the elements of civil aircraft operating cost, both direct and indirect [57].

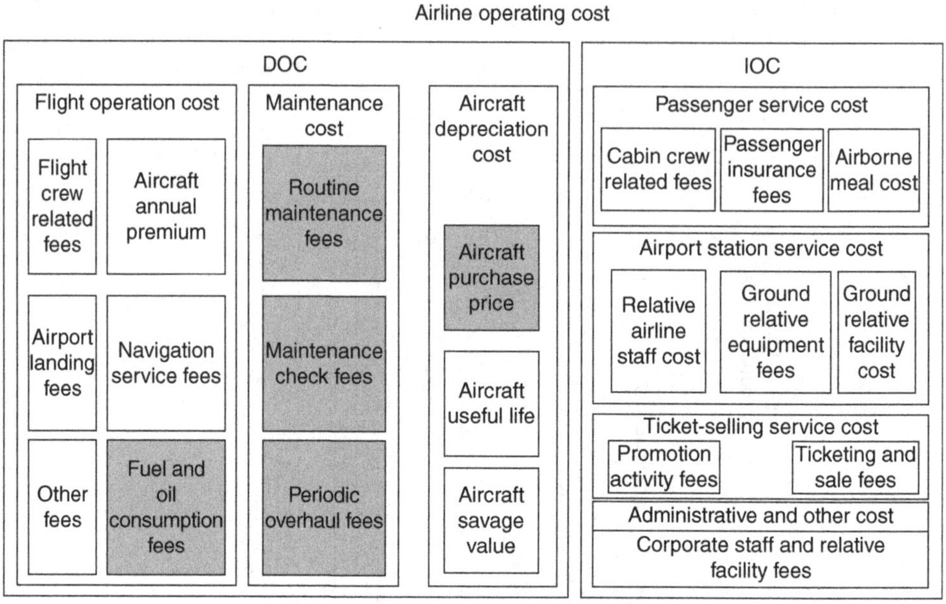

Airline operating cost

Fig. 8.6 Civil aircraft operating cost elements [57].

The improvement of reliability and maintainability will be discussed below, after consideration of military aircraft life-cycle costs.

Many manufacturers, airlines and government organizations have developed their own DOC formulae to compare different aircraft types. The method that is widely used in Europe is that of the Association of European Airlines [19].

8.4 MILITARY AIRCRAFT LIFE-CYCLE COSTS

The life-cycle costs of a military aircraft are analogous to the sum of both indirect and direct operating costs of civil transports. There is a surprising number of self-explanatory cost elements, as shown in Fig. 8.7.

Table 8.1 [15] shows a summary of the various elements of Royal Air Force expenditure during the mid-1980s. More recent information is unavailable, but recent trends towards 'contractorization' will have made significant differences. This process transfers much of the heavy maintenance to commercial companies, together with significant reductions in the employment of service personnel.

Again, the dominant effects of maintenance and reliability on the total costs of operating aircraft can be seen. The above costs cover the entire air force activities, not only those associated with flying aircraft. Table 8.2 gives a breakdown of the total life-cycle costs for a military aircraft [15].

Table 8.1   Royal Air Force expenditure in the mid-1980s

Item	Percentage of total
Development of new aircraft equipment Production	12%
Acquisition of new aircraft	11%
Spares and repairs	26%
Manpower and overheads	40%o
Fuel, etc	11%
Total	100%

Table 8.2   Life-cycle cost proportions

Life-cycle phase	Launch	Acquisition (development)	Initial support	Direct operations	Indirect operations	Total
Airframe (%)	13	10	3	4	0	30
Equipment (%)	4	8	5	9	0	26
Engine (%)	7	5	4	8	0	24
Support (%)	0	0	3	9	8	20
Total (%)	24	23	15	30	8	100

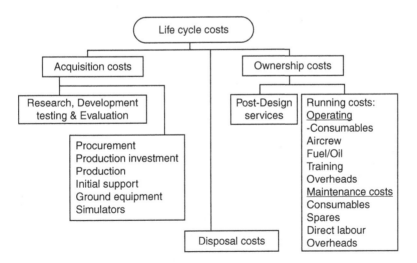

Fig. 8.7 Military aircraft life-cycle cost elements.

8.5 THE COSTS OF RELIABILITY AND MAINTAINABILITY

8.5.1 Civil Aircraft Unreliability Costs

The most dramatic form of unreliability is an aircraft crash with loss of life. It is impossible to quantify the value of a human life, although some attempts have been made to put cash values on lives. Aircraft crash losses are easier to quantify in terms of replacement and loss of revenue. Insurers have to cover human and material costs, and typically charge premiums of 1% of aircraft initial cost per annum. Figure 8.8 [58] shows a simplified view of the inter-relationships between safety, reliability and maintainability. It also shows design and analysis tools that may be used to improve performance in these vital areas.

The next item of unreliability cost is that of delays and cancellations. Figure 8.9 shows the proportions of all causes of delay. Figure 8.10 shows a typical breakdown of the systems causing technical delays. It can be seen that the aircraft structure has a relatively small effect on delay cost, but becomes very significant in older aircraft, when fatigue and corrosion can be particularly expensive.

The third major cost of unreliability is fault diagnosis, removal and repair or replacement of faulty components (which contribute to maintenance costs).

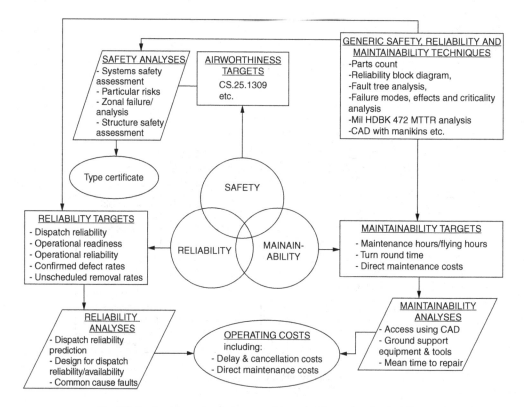

Fig. 8.8 Design interaction between safety, reliability and maintainability.

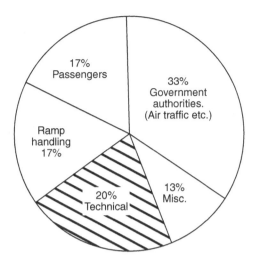

Fig. 8.9 Delay causes due to all reasons [19].

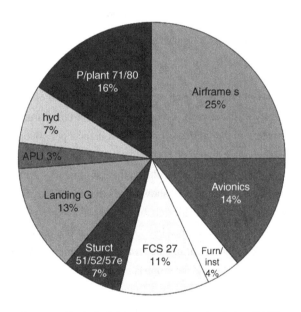

Fig. 8.10 Technical delay causes, short-haul aircraft [19].

8.5.2 Military Aircraft Unreliability Costs

Aircraft crashes are, again, costly in terms of life and material costs. Poor military aircraft reliability will severely limit aircraft availability and could, in extreme cases, lead to the loss of a military engagement. Most military aircraft reliability-related costs are, however, associated with increased maintenance costs. Failures will require the costly replacement or repair of components, together with the associated diagnostic and direct maintenance man-hour costs.

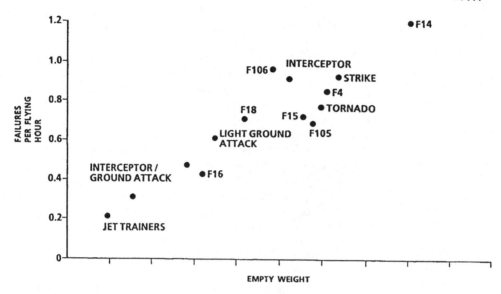

Fig. 8.11 Combat aircraft failure rates.

Figure 8.11 shows 1987 figures for failure rates against aircraft empty weight. The latter is a measure of the complexity of the aircraft. If one multiplied the failures per flying hour by 1000, one would have a figure close to the number of confirmed defects per 1000 h. That shows that by 1987 the Tornado was achieving a figure of 800 defects/1000 hours, compared with the target of 500. The curve shows a good correlation between weight and reliability. More recent, unclassified data are not available, but Figs. 8.11 and 8.12 show trends that are similar for more recent aircraft.

Figure 8.12 shows trends in specific reliability over a 25-year period. It is superficially depressing, because no apparent progress has been made. What has happened is that the aircraft performance, such as avionic capability or manoeuvrability, has improved dramatically, but at the cost of stagnation in terms of reliability. The emphasis has changed in more recent years. Specific reliability is defined as the number of failures per flying hour per unit of aircraft mass.

Figure 8.13 shows a drawing of the Cranfield TF-89 fighter project. This was designed to what was then known of the requirements, which led to the F-22 aircraft.

The aircraft is beingdesigned 'to be both lethal and survivable while penetrating high-threat air space. This will be achieved through a proper balance of increased speed and range, enhanced defensive avionics, reduced observables, and an emphasis on reliability and maintainability'. Through this 'proper balance', the goal becomes, quite simply, to 'double the sortie rate of the F–15', and 'do so with half the maintenance personnel required and twice the mean time between failures (MTBF)'. 'Reduced observables' means that the aircraft should have a low probability of being detected by radar, optical, acoustic or heat-seeking detectors.

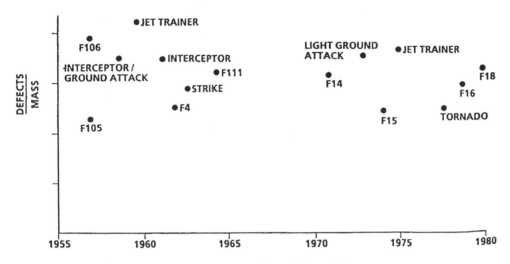

Fig. 8.12 Trends in specific reliability with time.

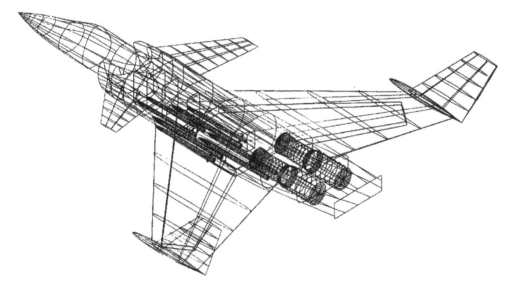

Fig. 8.13 Cranfield TF-89 Advanced Tactical Fighter Project.

The stringent reliability and maintainability targets had a significant impact on the conceptual design of the aircraft. The aircraft had a quickly removable weapons pannier for reduced turn-round times, more cooling for avionics, and other measures.

Uninhabited Aerial Vehicles (UAVs) are becoming increasingly important in both Civil and Military operations. An article in *National Defense Magazine* [20] stated:

Crashes and component failures are driving up the cost of unmanned air vehicles and availability for military operations, said a Pentagon report. . . .. The Defense Department 2002 roadmap confirms a mixed history of Class A mishaps, causing loss or severe damage to an aircraft. . ..While the F-16 manned aircraft was recently assessed at a mishap rate of 3,5 per 100,000 flight hours;

- The RQ-2A Pioneer had a mishap rate of 363 per 100,000 hours, though the figures for the RQ-2B declined to 139.

- The Predator RQ-1A had a mishap rate of 43, and the RQ-1B, 31.

In 2009, loss rates due to accidents were 10 per 100 000 hours for USAF Predators and 12.7 for Reaper UAVs [21].

8.5.3 Programmes to Reduce Reliability Costs

The cost of rectifying a fault is cheapest at the earliest possible stage in an aircraft's life-cycle. It is therefore important to spend money to get the reliability right in the first place.

There are several ways of attempting to do this:

(i) The setting of reasonable reliability targets (see Section 8.5.4 below).
(ii) Detailed reliability prediction analysis [fault trees and failure mode and effects analysis (FMEA)].
(iii) Design reviews to control reliability.
(iv) Extensive representative testing.
(v) Good reliability growth programme incorporating (ii) and (iv) with quick recognition and rectification of faults during development testing.
(vi) Good detail design, with good feedback of information from previous aircraft.
(vii) Use of well-proven components, possibly de-rated, so that the equipment is operated at a lower stress level than its theoretical maximum.
(viii) Attention to maintainability (see Section 8.5.5).

To do all these jobs well requires considerable manpower, equipment and the use of reliability engineers, which are costly. Some trade-off is therefore required between the costs of unreliability and the costs of a comprehensive reliability programme. Figure 8.14 Shows trends in reliability performance over an aircraft programme life-cycle.

8.5.4 Reliability Targets

8.5.4.1 Military Aircraft

As stated above, the setting of a reasonable target is of the utmost importance to the aircraft design, because too high targets can be crippling in terms of development costs, but too low targets limit aircraft availability. The two main types of targets are:

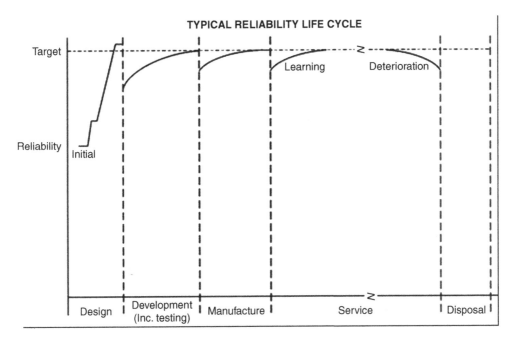

Fig. 8.14 Typical reliability life cycle [19].

(i) Mission reliability – This is the probability that the aircraft will be able to perform a given mission without any failures or defects that will have an operational effect. A modern combat aircraft has been recently designed to have an operational reliability of 0.95. The usual method of recording operational defects is to monitor the whole aircraft fleet over a number of months and produce an operational defect rate. This is defined as:

(No. of operational defects for the fleet in the period/Fleet flying hours) × 1000

Operational defects cover such things as accidents, mission cancelled, mission partial failure, flight safety hazard, etc. Thus the target mission reliability is 50 operational defects per 1000 h. A BAE SYSTEMS Hawk aircraft brochure quoted a target of 0.98 for mission reliability, or 20 operational defects per 1000 h.

The EMBRAER KC-390 military transport aircraft mission reliability target was given as 95% [22].

(ii) Aircraft confirmed defect rate – Air Forces have comprehensive reliability data recording systems. Records are maintained of all defects reported on aircraft, however, not all defects are confirmed as genuine defects because some of the reports may have been due to poor fault diagnosis.

8.5.4.2 Civil Aircraft Targets and Reliability Monitoring

(i) Safety targets – Aircraft must be designed to be extremely safe, but, *absolute safety* cannot be achieved. The Airworthiness Authorities have examined the overall aircraft

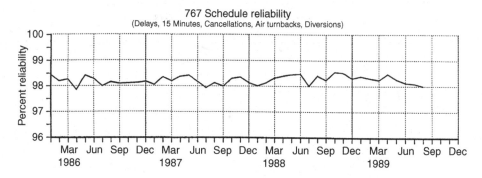

Fig. 8.15 Boeing 767 Schedule reliability.

accident rates and have set a target of 1 fatal accident in 10^7 flying hours. This target is then subdivided into the various systems of the aircraft, for example auto-landing failures must not exceed 1 in 10^9.

(ii) Delay rates – If the aircraft fails to depart on time for technical reasons it is counted as a technical delay (on large aircraft it is not recorded unless the delay is more than 15 minutes). Delays are attributed to the system that has caused the problem and delay rates are calculated and displayed as shown in Figure 8.15. This is a useful management tool that can be used to monitor reliability.

(iii) Technical log entries – There is a legal requirement for flight-crew members to report all safety or reliability problems discovered during a flight. These may be converted into technical log entry rates and displayed in a similar manner to delays as shown in Figure 8.16. They are sometimes called pireps (pilot reports).

(iv) Maintenance reliability unscheduled removal rates – This is a component-orientated measure of reliability. If a component is suspected of being defective it is removed from the aircraft and replaced by a new item. Many hundreds of components are monitored, and tables such as Figure 8.17 are published at monthly intervals. This highlights troublesome components, and can also be used for spares stock-control purposes. Information from this source further provides information for safety and delay rate analyses.

8.5.5 Factors Affecting Maintenance Costs

The magnitude of maintenance costs has already been mentioned, and is closely linked to the unreliability of the aircraft. An aircraft with low operational costs will need to have high reliability. Even the most reliable components, however, will fail and will need to be replaced. Good design for maintenance will minimize such costs. Figure 8.18 shows an overall plan for maintainability (ease of maintenance). Not all of these features are within the control of the designer, but he or she must be aware of them.

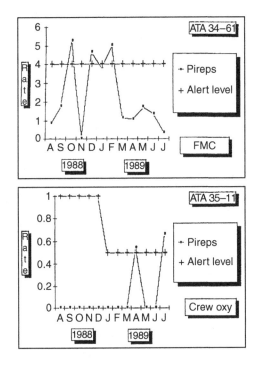

Fig. 8.16 Pilot report (pirep) rates for Boeing 767 systems.

Embraer	Emb-120 Brasilia - Component unscheduled removals	August 89	
		Section 4	Page 4

24 - Electrical power

Nomenclature	Part number	Month	Last 3 M		Last 12 M		Accumulated	
		Qty	Qty	Mtbur	Qty	Mtbur	Qty	Mtbur
Battery	167556	1	1	53, 609	13	12, 709	106	3, 522
Starter generator	23080–013	13	47	3, 387	143	3, 364	362	2, 946
Dc Generator	30081–001	9	22	4, 874	38	8, 696	94	7, 943
Emergency battery	501–1228–04	3	9	5, 957	26	6, 355	51	7, 320
Generator control unit	51539–014 A	18	54	1, 986	173	1, 910	253	2, 951
Generator control unit	51549–001 A	1	7	15, 317	14	23, 603	39	19, 144
Volt-Ammeter	522541	3	6	17, 870	28	11, 801	55	13, 575
Temperature monitor	522545	2	4	13, 402	18	9, 179	52	7, 179
Static inverter	PC–251–1C	2.6	69	3, 108	100	6, 609	130	11, 486
Rccb	SM600BA75A1				3	110, 147	5	149, 324
Rccb	SM600BA60N1	2	2	53, 609	6	55, 073	9	82, 958

Fig. 8.17 Electrical power system unscheduled removals.

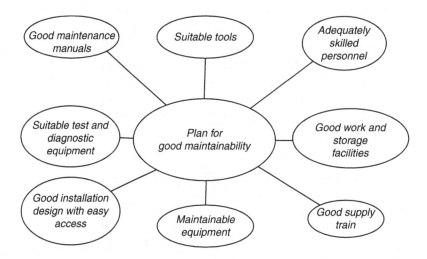

Fig. 8.18 The main factors affecting maintenance.

8.5.5.1 Tools

The designer must be aware of what tools are available in either the hangar or flight operations areas and how they are used, or misused. Minimizing the number of tools used must be a priority.

8.5.5.2 Personnel

A designer must know the quality of maintenance personnel. Are they semi-skilled conscripts or highly professional mechanics? What environment are they working in? Naval designers have a saying, 'Is it Jack-proof?'.

8.5.5.3 Work and Storage Facilities

The work and storage facilities that are available dictate the type of structural construction methods, and the size and complexity of equipment modules, number of spares required, etc.

8.5.5.4 Supply

The quality of the supply train, again, is a factor in spares holding and turn-around times. If supply lines between equipment stores and operating aircraft are poor, it might be better to attempt field repairs.

8.5.5.5 Maintainable Equipment

Such equipment should be designed for easy fault diagnosis, removal and rectification. This is a vast subject so only a few examples will be shown. Figure 8.19 shows a modular aircraft engine. This design means that engines can be dissembled *in situ* and modules rather than engines can be

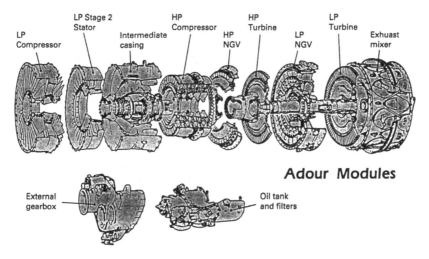

Fig. 8.19 Modular engine.

changed and refurbished. As engines are such a high-cost maintenance item, this is important. Other features are access points for boroscopes, magnetic plugs etc.

Avionic equipment is often mounted in standard Austin Trumbell Radio (ATR) boxes, which are easily removable and interchangeable.

Figure 8.20 shows some examples of good maintainability features.

8.5.5.6 Installation and Accessibility

An experienced maintenance engineer has made the following comments about access:

Accessibility is always a primary problem in design for effective maintenance, and the cleverest trick of the good designer is to get a 'quart into pint pot' and be able to get it out again. When you consider accessibility in a design, it must not be forgotten that the vehicle may perform its tactical function operating out of ill-equipped and unprepared forward bases during the exigencies of wartime. The majority of maintenance functions should, where practicable, be accessible from the ground, or at the worst, using a relatively low work stand, or in an emergency, up-turned packing cases. Access holes must be large and easily opened, and should permit the maximum number of men working simultaneously which are normally required for the performance of maintenance tasks on the affected systems.

Figure 8.21 shows the great lengths taken by the designers of the BAE SYSTEMS Hawk to ensure good access for maintenance.

8.5.5.7 Test Equipment

There are many types of built-in test equipment (BITE), but all of them are designed into the component to indicate its condition. The simplest type is the 'dolls-eye' indicator, which shows failures on control panels.

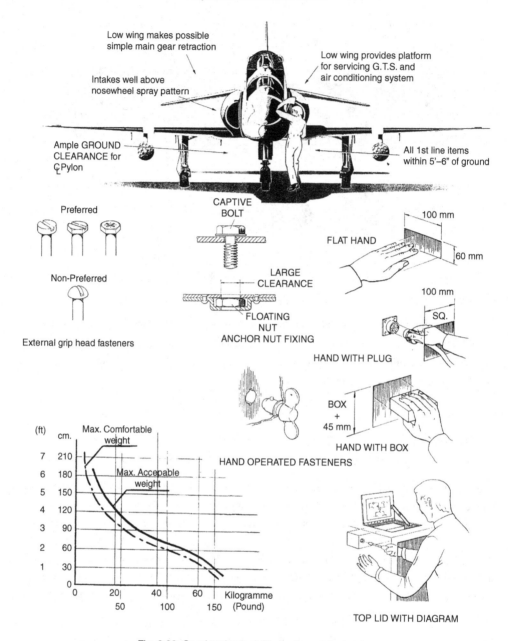

Fig. 8.20 Good maintainability equipment features.

As with all test equipment, BITE ought, ideally, to be an order of magnitude more reliable than the system of which it is part, in order to minimize the incidence of false alarms, false diagnoses, incorrect alignments or even failure to register a fault condition. All these events will either prolong down time or reduce the reliability of the system, and therefore built-in test equipment of poor reliability can grossly reduce the system effectiveness.

⊓△⫣⫢⨏⌿⪆ Access Areas

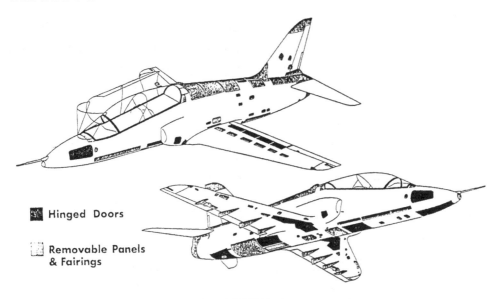

Hinged Doors

Removable Panels
& Fairings

Fig. 8.21 BAE SYSTEMS Hawk access panels.

The introduction of digital avionic systems and microprocessors has greatly improved the reliability and capabilities of BITE systems.

Many complex electronic systems are built up of easily removable modules. If there are many of these modules or if quick turn-round is required, the expense of automatic test equipment (ATE) may be justified. The ATE is a complex piece of equipment that is connected up to the suspect module or system. It then performs an automatic test sequence and rejects defective equipment, which may then be overhauled.

8.5.5.8 Maintenance Manuals

The maintenance review board (MRB) process was an important innovation, introduced during the design of the Boeing 747. Subsequent US wide-body aircraft and European aircraft such as the Concorde and A300B have also used this system, or modifications of it.

The maintenance planning for a new aircraft is done by a maintenance steering committee. Representatives from the following organizations sit on the committee: the airframe manufacturer; operators who will buy the aircraft; the powerplant manufacturer; and the airworthiness authority. Working groups act as subcommittees to the steering group and work in specialist areas. Component manufacturers are included in these groups, which might specialize in such areas as avionics, mechanical systems, etc.

These groups work through the logical decision-making processes [23] to determine scheduled and unscheduled maintenance procedures. This system considers the safety, reliability

Fig. 8.22 Maintenance access check using a CAD manikin maintainer during the Cranfield U-3 UAV design project.

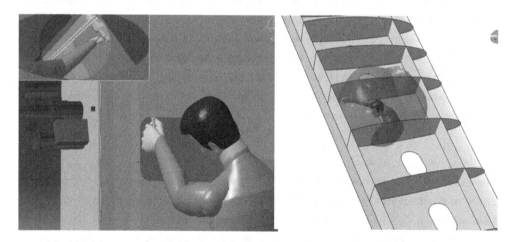

Fig. 8.23 Maintenance access for actuator and wing box.

and cost aspects of scheduled maintenance by the use of decision trees. The process also throws light on to the unscheduled maintenance procedures.

The MRB produces a report that forms the basis of an approved maintenance schedule for the aircraft. This leads to the writing of maintenance manuals that control the way in which aircraft are maintained.

8.5.6 Maintenance Targets

A quantitative target is always useful to encourage a designer to improve maintainability. This requires extensive knowledge of the performance of previous products so that sensible targets are set. Civil aircraft operators often demand direct maintenance cost guarantees for both the aircraft and its engines.

Military aircraft targets are quoted as the number of maintenance man-hours per flying hour. The BAE SYSTEMS Hawk is a good example of an aircraft designed for good maintenance, and its performance is as follows:

Overall maintenance = 3.8 man-hours/flying hour.

This figure includes all maintenance, including flight line and is about half that of the Gnat and Hunter aircraft that preceded it.

Flight line maintenance achieved figures are:

$$\left. \begin{array}{l} \text{Pre-flight} = 1 \text{ man in } 13 \text{ mins} \\ \text{Turn-round} = 1 \text{ man in } 9 \text{ mins} \\ \text{Post-flight} = 3 \text{ men in } 11 \text{ mins} \\ \text{Re-arming} = 4 \text{ men in } 10 \text{ mins} \end{array} \right\} \text{in parallel}$$

It is very difficult to quantify these figures during the initial design stages, but comparisons can be made with empirical figures obtained from current aircraft. Maintainability engineers can advise designers of good practices, and mock-ups can be used in the relatively early stages of design. Maintenance trials can be performed with mock-ups where it is still possible to change the design to improve matters. Maintenance checks are then made at subsequent stages of development, but the further on the development process, the more difficult it becomes to make changes. Computer-aided design (CAD) can be a useful tool to check accessibility for maintenance purposes. CAD full-scale models have been made that can represent the shapes and strengths of the complete range of maintenance personnel. These early checks can be made in what are termed 'electronic mock-ups' (see Figs. 8.22 and 8.23).

The use of quantified maintenance targets, and the other means described above have the potential to significantly reduce maintenance costs.

CHAPTER 9

WHAT HELP CAN I GET?: BIBLIOGRAPHY AND COMPUTER-AIDED DESIGN

The previous chapters have shown the complexity of aircraft, and their constituent parts. Their design is a daunting, but potentially satisfying task. Designers need all the help that they can get to achieve successful results.

The usual starting point is a sound aeronautical education. Design is learnt by a combination of theoretical education and design experience at university and/or in industry. It must be stressed that design is much more than analysis. It is a creative process involving the synthesis of knowledge from many disciplines, monitored by qualitative and quantitative checks. These should assess the value of the necessary design compromises and lead to optimum designs. Advice from, and work alongside, experienced designers is an invaluable part of the education process. People learn how to design by actually doing it! Less-experienced design organizations can be helped by specialist consultants. Such help is, however, not always available, but much design knowledge and experience has been encapsulated in the publications and programs described later in this chapter.

Many individuals and organizations publish reports, papers and memoranda, many of which contain relevant data. Modern libraries use on-line computerized data bases that sort and codify these date sources on the basis of key words. A description of such reports is beyond the scope of this book. The most exciting new development has been the advent of the internet. There is a bewildering array of data, with much of interest to aircraft designers. Patience and good net-surfing skills are required.

Any bibliography is a matter of personal choice. The books, magazines and programs listed below have been found by the author to be extremely useful and are reasonably easily available. Excluded data sources are not necessarily less valuable than those listed, but are not available to the author.

The aircraft industry has seen a massive growth in the power and user-friendliness of computer aids. The latter part of this chapter gives an overview of a range of computerized design aids.

9.1 AIRCRAFT DESIGN BIBLIOGRAPHY

9.1.1 Conceptual Design

HOWE, D. *Aircraft Conceptual Design Synthesis*. Professional Engineering Publishing Limited. London. 2000. An excellent book on conceptual design by my predecessor.

OBERT, E. *Aerodynamic Design of Transport Aircraft*. IOS Press (Delft University), Amsterdam, the Netherlands. 2009. Written by the former Chief Aerodynamicist of Fokker, with fascinating detail, not available elsewhere in the public domain.

NICHOLAI, L. M., CARICHNER, G.E. *Fundamentals of Aircraft and Airship Design, Volume 1: Aicraft Design*. AIAA Education series. 2010. This is a very large book, with a mine of useful, up-to-date data and design tools.

RAYMER, D. P. *Aircraft Design: A Conceptual Approach*. 2nd edn. American Institute of Aeronautics and Astronautics, 1993. This book has an emphasis on combat aircraft and has associated PC-based design software.

ROSKAM J. *Airplane Design Series*. Parts I–VIII. Roskam Aviation and Engineering Corporation, 1989. This is a large series with associated software for workstations.

STINTON, D. *The Design of the Aeroplane*. Granada, London, 1983. This book is particularly good for the design of light aircraft.

THURSTON, D. B. *Design for Flying*. McGraw-Hill, New York, 1978. This book has a useful section on seaplanes.

TORENBEEK, E. *Synthesis of Subsonic Aircraft Design*. (Martinus Nijhof), Delft University Press, 1982. This is a widely-used text, particularly appropriate for civil transport aircraft.

WHITFORD, R. *Design for Air Combat*. Jane's Publishing Co, Ltd., London, 1987.

WILKINSON, R. *Aircraft Structures and Systems*. 3rd edn. MechAero Publishing, St. Albans, UK. 2011. A well-illustrated introductory book covering civil and military aircraft structure and systems.

9.1.2 Aerodynamics Books

ABBOTT, I., VON DOENHOFF, A. *Theory of Wing Sections*. Dover Publishing Co. New York, 1959.

BARNARD, R. H., PHILPOTT, D. R. *Aircraft Flight*. Longmans, London, 1989.

BABISTER, A. W. *Aircraft Dynamic Stability and Response*. Pergamon Press, Oxford, 1980.

BRAMWELL, A. W. S. *Helicopter Dynamics*. Arnold, London, 1976.

BISPLINGHOFF, R. L., ASHLEY, H. *Principles of Aeroelasticity* Dover Publishing Co., New York, 1975.

HOERNER, S. E. *Fluid Dynamic Drag.* Hoerner Fluid Dynamics, PO Box 342 Bricktown NJ 08723, USA.

HOERNER, S. F. and **BORST, H. V.**, *Fluid Dynamic Lift.* Hoerner Fluid Dynamics, PO Box 342, Bricktown, NJ 08723, USA, 1975.

HOUGHTON, E. L, CARPENTER, P. W. *Aerodynamics for Engineering Students.* 4th edn. Edward Arnold, London, 1993.

KUCHEMANN, D. *The Aerodynamic Design of Aircraft.* Pergamon Press, Oxford 1978.

MAIR, W. A., BIRDSALL D. L. *Aircraft Performance.* Cambridge Aerospace Series, Cambridge University Press, UK.

SEDDON, J., GOLDSMITH, E. L. *Intake Aerodynamics.* AIAA, Waldorf, MD, USA, 1986.

VINH, N. X. *Performance of High Performance Aircraft.* Cambridge Aerospace Series, Cambridge University Press, UK.

9.1.3 Structures Books

BRUHN, E. F. *Analysis and Design of Flight Vehicle Structures.* Cincinnati, Ohio, Tristate Offset Co. (This may be difficult to obtain.)

CUTLER, J. *Understanding Aircraft Structures.* Granada, London, 1981.

CHUN-YOUNG NIU, M. *Airframe Structural Design.* Conmilit, USA, 1990.

MEGSON, T. H. G. *Aircraft Structures for Engineering Students.* Edward Arnold, London, 1990.

TIMOSHENKO, S. P., GERE, J. M. *Mechanics of Materials.* Van Nostrand Reinhold, New York, 1969.

TSAI, S. W., HAHN, H. T. *Introduction to Composite Materials.* Westport, Conn., Technical Pub. Co., 1980 (when available).

9.1.4 Systems and General Books

ANON *AIAA Aerospace Design Engineers Guide.* 3rd edn. AIAA, Waldorf, MD, USA, 1993.

BALL, R. E. *The Fundamentals of Aircraft Combat Survivability: Analysis and Design.* AIAA, Waldorf, MD, USA 1985.

CURREY, N. S. *Aircraft Landing Gear Design: Principles and Practices.* AIAA, Waldorf, MD, 1988.

Jane's *All the World's Aircraft* edited by John W. R. Taylor. Published annually by Jane's Publishing, London and New York. This is an invaluable, but expensive, book. It gives extensive details of all aircraft in current production. *It is possible to buy it on a ROM disk.*

MATTINGLEY, J. D., HEISER, W. H., DALEY, D. H. *Aircraft Engine Design.* AIAA, Waldorf, MD, USA, 1987.

MOIR, I., SEABRIDGE, A. *Aircraft Systems.* Longman Scientific and Technical, UK. 1992.

NEESE, W. A. *Aircraft Hydraulic Systems.* Pitman Publishing Ltd, London, UK, 1984.

PALLETT, E. H. J. *Aircraft Electrical Systems.* 3rd edn. Longman Scientific and Technical, UK, 1987.

ROLF, J. M., STAPLES, K. J. *Flight Simulation.* Cambridge Aerospace Series, Cambridge University Press, Cambridge, UK.

SMITH, M. J. T. *Aircraft Noise.* Cambridge Aerospace Series, Cambridge University Press, Cambridge, UK.

9.1.5 Magazines and Journals

Aerospace, The Royal Aeronautical Society, London, UK, monthly.

Aerospace America, AIAA, USA, monthly.

Aviation Week and Space Technology, USA, weekly.

Flight International, UK, weekly.

Journal of Aircraft, AIAA, USA, monthly.

Interavia, Geneva, Switzerland, monthly.

9.2 DATA SHEETS, AIRWORTHINESS REQUIREMENTS AND ENVIRONMENTAL TARGETS

Data sheets contain vast amounts of empirical and theoretical data, amassed over decades. The British ESDU series evolved from the Royal Aeronautical Society data sheets and cover a wide range of topics. The main aeronautical series are listed later. The USAF DATCOM series are particularly concerned with aerodynamics, and stability and control. ESDU uses British aerodynamic notation, whilst DATCOM uses American terminology.

Aircraft can only be flown legally if they have been designed, certificated, manufactured and operated to recognized standards. The starting point is compliance with Airworthiness Requirements, which will be summarized in Section 9.2.4.

Aircraft-related environmental targets are being developed, some of which are shown in Section 9.2.5.

9.2.1 ESDU Data Sheets

These listed below are particularly useful to aircraft designers, but there are other series. These may be obtained from:

IHS ESDU,123, Houndsditch, London, EC3A 7BX, UK
Website: https://esdu.com
ESDU Aerodynamics Series
Section 1: Organisational Documents
Section 2: Properties of Gases
Section 3: Isentropic Flow and Shock Waves
Section 4: Properties of the Atmosphere
Section 6: Wind Speeds
Section 7: Aerofoils and Wings – General
Section 8: Aerofoils at subcritical Speeds – Pressure Distribution, Lift, Pitching Moment, Aerodynamic Centre
Section 9: Aerofoils at Subcritical Speeds – Drag
Section 10: Aerofoils at Supersonic Speeds – Pressure Distribution, Lift, Pitching Moment, Drag
Section 11: Critical Mach Number and Pressure Coefficient
Section 12: Flat plates – Boundary layers, Skin Friction and Surface Roughness
Section 13: Wings – Lift, Pitching Moment, Aerodynamic Centre, Spanwise Loading
Section 14: Wings – Drag
Section 15: Bodies – General
Section 16: Bodies – Drag
Section 17: Bodies – Pressure Distribution, Normal Force, Pitching Moment, Centre of Pressure
Section 18: Wing-Body Combinations – Lift, Normal Force, Pitching Moment, Aerodynamic Centre, Upwash
Section 19: Wing-Body Combinations – Drag
Sections 20–27: Controls and Flaps

Section 28: Excrescence Drag

Section 29: Cavity Drag

Section 30: Undercarriage Drag

Section 31: Canopy Drag

Section 32: Cavity Aerodynamics and Aero-Acoustics

Section 33: Cavity Adverse Unsteady Flow Alleviation

Section 34: Internal Flow Systems – Ducts

Section 35: Internal Flow Systems – Nacelles, Intakes and Nozzles

Section 36: Powerplant/Airframe Interactions – Propeller Powered Aircraft

Section 37: Powerplant/Airframe Interactions – Jet Powered Aircraft

Sections 38–44: Stability of Aircraft

Section 45: Unsteady Aerodynamics

Section 46: Parachute Aerodynamics

Section 47: Bluff Bodies and Structures – Mean Forces

Section 48: Bluff Bodies and Structures – Fluctuating Forces and Response

Section 49: Aerodynamic Heating and Heat Transfer

Section 50: Wind-Tunnel Corrections

Composites

Performance

Stress and Strength

Structures

Transonic Aerodynamics

Fatigue

ESDU Pac computer Programs

9.2.2 USAF Stability and Control DATCOM

This was originally published in October 1960 at Wright Patterson Air Force Base, USA. It has been subsequently revised. The sections are:

Section 1   Guide to DATCOM and methods summary
Section 2   General information
Section 3   Reserved for future use
Section 4   Characteristics of angle of attack
    4.1   Wings at incidence
    4.1.1   Lift curve shape
    4.1.2   Section – Pitching moments
    4.1.3   Wing lift (3 dimensional)
    4.1.4   Wing pitching moment

9.2.3 Digital DATCOM

A more recent development has been the free availability of the above printed DATCOM in electronic form, termed digital DATCOM. It is freely available on http://www.pdas.com/datcomres.html. It also contains very useful computer programs to calculate static and dynamic stability, high lift and control characteristics, together with a trim option.

9.2.4 Airworthiness Requirements

9.2.4.1 Civil Aircraft European Airworthiness Requirements: EASA
(Similar categories apply to US FAR requirements, and for many other countries.)

CS-22 Sailplanes and Powered Sailplanes

CS-23 Light Aeroplanes (<5700kg) + commuter (<8618kg)

CS-25 Large Aeroplanes

9.2.4.2 Military Aircraft Airworthiness Requirements

The *UK Def-Stan 00–970*, published the Ministry of Defence, has the following categories:

I – Small light aeroplanes

II – Medium weight, low and medium manoeuvrability

III – Large, heavy, low and medium manoeuvrability

IV – High manoeuvrability

The USA Department of Defense publishes many *Mil Standards*.

9.2.5 Aircraft Environmental Targets

There are many targets, the most important being those produced by the ICAO (International Civil Aviation Organization), particularly for noise and airports. See www.icao.int.

The most recent European Environmental targets came from the European Union Report - Flightpath 2050:

"1. In 2050 technologies and procedures available allow a 75% reduction in CO_2 emissions per passenger kilometre and a 90% reduction in NOx emissions. The perceived noise emission of flying aircraft is reduced by 65%. These are relative to the capabilities of typical new aircraft in 2000.

2. Aircraft movements are emission-free when taxiing.

3. Air vehicles are designed and manufactured to be recyclable.

4. Europe is established as a centre of excellence on sustainable alternative fuels, including those for aviation, based on a strong European energy policy.

5. Europe is at the forefront of atmospheric research and takes the lead in the formulation of a prioritised environmental action plan and establishment of global environmental standards."

9.3 COMPUTER DESIGN TOOLS

The aeronautical industry was one of the pioneers in the use of computers and has maintained a position of pre-eminence in the field. They are currently used across the whole spectrum of conceptual design, analysis, detail design, testing, manufacture and operations. Their importance is such that all aeronautical engineers should be aware of their use. This section will describe some

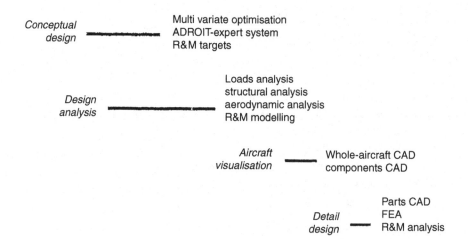

Fig. 9.1 Design aid categories.

of the computer tools that are being developed at Cranfield for aircraft design. They are typical of such tools that have been developed at other universities, in industry and research establishments, and give an indication of the computer aids available to designers.

Cranfield's philosophy is to use computers as a natural part of its activities. Students are given 'hands-on' experience in as many computer applications as possible, so that they will be well equipped to make significant contributions in their subsequent activities. As the range of work is very wide, it is intended to limit the description of activities to work in aircraft conceptual design, analysis, aircraft visualization and detail design (Fig. 9.1).

9.3.1 Conceptual Design Programs

9.3.1.1 Multi-Variate Optimization Programs

A design synthesis program has been developed for *canard* combat aircraft. The system synthesizes the shape of the aircraft in terms of pilot position, armament, wing size etc. It calculates lift, drag, mass and performance. It may be used in the design process, or to quickly assess existing designs. The PhD thesis of Serghides [25] describes the start of this work. The thesis synthesis has now been linked to an optimizer and graphical program and Fig. 9.2 shows an example of the program output.

Modifications have been introduced to cater for configurations that include novel propulsion systems required for supersonic STOVL layouts (short take-off, vertical landing) and 'stealth' aircraft.

A similar program has been developed for the conceptual design of unmanned aircraft (UMA). It will synthesize and optimize the configuration of a wide range of subsonic UMA.

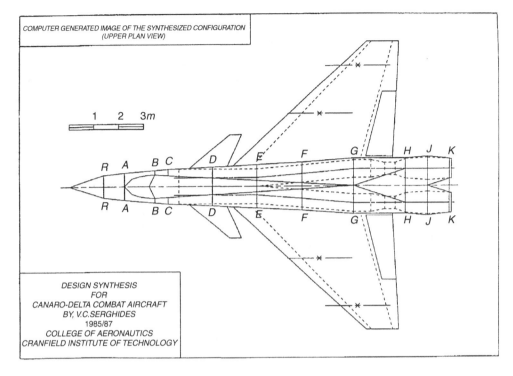

COMPUTER GENERATED IMAGE OF THE SYNTHESIZED CONFIGURATION
(UPPER PLAN VIEW)

DESIGN SYNTHESIS
FOR
CANARO-DELTA COMBAT AIRCRAFT
BY, V.C.SERGHIDES
1985/87
COLLEGE OF AERONAUTICS
CRANFIELD INSTITUTE OF TECHNOLOGY

Fig. 9.2 Graphics plot of canard design.

Configurations include podded and integral jet engines and a number of tractor and pusher propeller layouts.

Work has also been performed on the use of synthesis and optimization of the conceptual design of supersonic transport aircraft, laminar-flow aircraft and commuter aircraft.

9.3.1.2 An Example of an 'Expert System': The ADROIT Aircraft Design System
The initial work, which started in 1984, was for an early investigation into what were termed 'expert systems' in which human design knowledge and logic are encapsulated in computer programs. This is a very simple example of what has become a major area of computer science.

Since the aircraft design task was very large, it was decided to take a representative piece of the design, that of the wing, in order to 'learn' the methodology used in designing complex artifacts such as aircraft. The ADROIT (aircraft design by regulation of independent tasks) program is a prototype that designs an aircraft wing for a subsonic airliner. It selects a two-dimensional aerofoil section from a choice of possibilities and evaluates a range of suitable sweep angles according to the aircraft specification. Within ADROIT, a controller monitors the execution of the different design steps and allows the user to re-evaluate aerodynamic, structural and layout characteristics to give a solution that satisfies these criteria. More recent work has produced a similar system to that of the wing for the design of the fuselage interior and exterior for wide-body, narrow-body and regional transport aircraft.

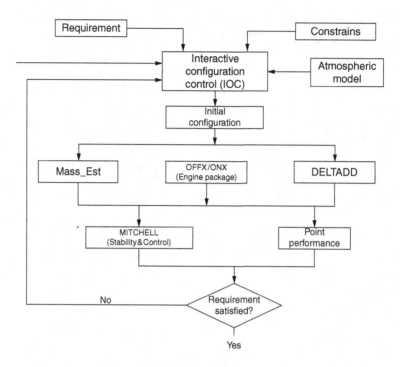

Fig. 9.3 Design analysis programs.

The most challenging task was the development of the configuration module. This is the heart of the design process and the program advises the user as to the best aircraft layout to meet a given specification. Such matters as engine type, number and position, wing and tail position, and undercarriage layout are assessed. The consequences of decisions are then examined and advice given to the user. There is a powerful explanation facility that shows the logic used in design-making. The configuration module is largely qualitative, but links have been made to some of the analysis programs described below.

9.3.2 Project Design Analysis Programs

The first aerospace applications of computers were in the analysis field – for 'number-crunching'. This is still an a important area and Cranneld has developed a large number of programs that are used in aircraft design, some of which are shown below.

Figure 9.3 shows the architecture of a partially completed suite of programs that are being developed for use in conceptual design analyses. This is inevitably a long-term process, but each program is useful in its own right and is immediately available. The aim is to teach good programming practices and insist on well-commented programs, augmented by user and programmer guides. The theoretical basis of programs must be stated and, if possible, manual worked examples given. Validation is also performed against known 'real' aircraft properties, such as performance and weight.

Particular attention has been given to the development of aircraft reliability, maintainability and mission-readiness programs. These were intended for use during the early design phases for the prediction of aircraft system performance and the setting of reliability and maintainability targets. This work was summarized by the author [26] and further work has led to computer tools that enable the user to optimize the use of bleed air to improve avionic system reliability [27].

9.3.3 Aircraft Visualization

Many types of computer-aided design (CAD) systems are available but Cranfield University has standardized on the CATIA 3-D system. This is typical of a wide range of classic CAD systems, which are described in more detail in Section 9.4. Its operation is demonstrated to all students who are then given hands-on experience. All design students in the aerospace vehicle design course are given the option of using the system for design development, and drawing production for their theses. Most students use the system extensively in part creation, geometry, interfaces, ergonomics, calculations and detail drawing. Examples will be given later. One particularly useful application was a check of engine removal for a military aircraft project. The fuselage was constructed as an 'electronic mock-up', with the advantage that any scale could be used.

Figure 9.4 shows an example of a fuselage interior produced by students, but more examples will be given in Section 9.5, below.

9.3.4 Programs Developed for Use in Preliminary/Detail Design

Many programs have been developed to help with analysis at the detail design stage. They are primarily concerned with structural and mechanical design, as that is the primary activity at the detail design stage. Topics include undercarriage modelling, flap configuration, fatigue analysis, fuselage and wing structural analysis, control surface structure, and aerodynamic panel analysis.

Further programs cover maintainability prediction, composite analysis and aircraft loading analysis. A number of proprietary programs are also used. The CAD packages have been mentioned, but extensive use is also made of finite element structural analysis and computational fluid dynamics packages. CAD is also used at the detail stage to produce 'electronic' drawings. The analysis facilities of CAD are used for such activities as the determination of fuel tank volumes.

Students are taught how to use these programs and are then encouraged to use them on design projects. It is felt that it is important that students are taught the limitations of the programs that they use. This is to guard against the tendency for people to use 'off-the-shelf' programs, with little knowledge of what analysis is being performed within them.

Recent work has included the development of an integrated CAD/CAM method for the design and construction of a wing component. The program first performs the conceptual design of an airliner wing on the CAD system, followed by parametric design of ribs. The system then was

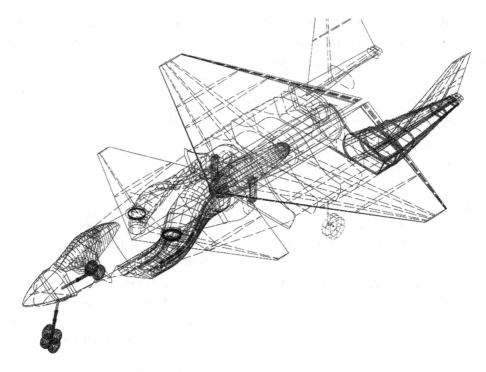

Fig. 9.4 CAD model of a fighter aircraft interior.

used to define CAM activity, which then led to the production of the component on a numerically controlled (NC) machine. This process is widely used by the aircraft industry as a part of the process termed concurrent engineering.

9.4 THE INTEGRATION OF COMPUTER TOOLS AS PART OF CONCURRENT ENGINEERING

Recent years have seen an explosion in the use of CAD across many industries. This process has been led by the aerospace industry; indeed many of the currently available systems were developed by aerospace companies. Several CAD systems are used in each aerospace factory for different purposes. CAD is much more than electronic draughting.

Fuselage, wing and configuration design is aided by conventional CAD systems by improving visualization of the design. It must be stressed that these systems are *aids* to the designer. Computers have also been used by decades for such analyses as mass and drag, performance, stability and control, and cost estimation; these are performed by dedicated programs *not* CAD systems.

During the preliminary and detail design stages, enough is known about the aircraft to define its shape more exactly. Accurate analyses are possible using computational fluid mechanics for solving aerodynamic problems. When the shape has been refined by such a process (and

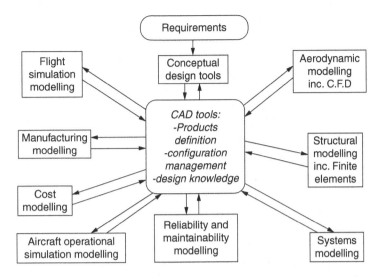

Fig. 9.5 Concurrent engineering tools.

wind-tunnel testing), more detailed design is started. Here CAD models define surfaces and interfaces. Components such as spars and frames are initially designed. There may be interfaces with finite element programs for structural analysis. It is after this stage that detailed designs and drawings are produced. Most CAD systems include, or interface with, NC machine systems such that no conventional drawings are required. Figure 9.5 shows how computer tools may interface with each other. Many companies still use drawings, but increasing numbers are produced by CAD systems. Computer design was used for 35% of the Boeing 767 structure, and about 40% for the 757. This accounted for some 7000 drawing sheets per aircraft. A more recent report stated that the same company had achieved 100% 'digital product definition' for the Boeing 777 project. Other large aerospace companies are progressing in a similar fashion.

The whole process of integration has led to the term 'concurrent engineering'. The design activities are integrated by the use of multi-disciplinary 'design-build teams'. Groups of engineers work on a design and introduce skills in design integration, structural design, aerodynamics, systems and manufacturing. In some companies, representatives of the aircraft operators are included. This ensures that the design will be able to be built efficiently in the factory, and maintained and flown successfully when it reaches service. Such an integrated process improves build quality and reduces the design/manufacturing timescale, relative to more conventional methods.

9.5 CLASSIC COMPUTER-AIDED DESIGN SYSTEMS

This section will concentrate on two-dimensional draughting, 3-D wire frame models and solid modelling. Many examples will be given of the use of such systems.

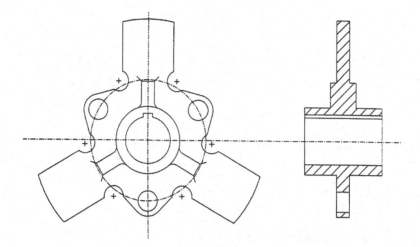

Fig. 9.6 Two-dimensional CAD image of a machined part.

9.5.1 Two-Dimensional Draughting

There are many CAD systems that will perform this function, but they all have common features.

9.5.1.1 Geometric Entities

CAD systems make use of simple entities such as points, lines and arcs. More sophisticated systems allow the construction of conic sections, polygons and splines. The latter are particularly useful for aerofoils and complex shapes.

9.5.1.2 Transformations

The time-saving transformation functions allow collections of entities to be moved, copied and duplicated several times with any of the following transformations:

Translation – linear motion, displacement.
Scaling – up or down.
Rotation – about a line or point.
Mirroring – through a plane or line.
Reposition – into another co-ordinate system.

These are all completely general 3-D transformations. Reposition is a powerful and useful transformation that involves both translation and re-orientation. The transformation functions also allow the automatic creation of rectangular and circular arrays of entities.

Figure 9.6 shows a part constructed by the author and shows the use of rotation and mirroring.

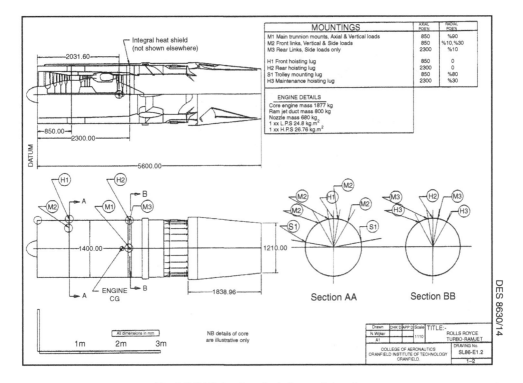

Fig. 9.7 CAD drawing of a turbo-ramjet engine.

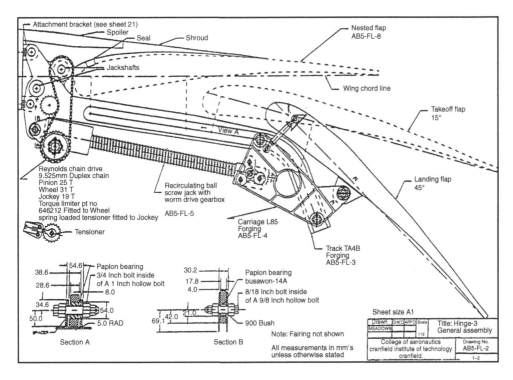

Fig. 9.8 Wing flaps for an airliner project.

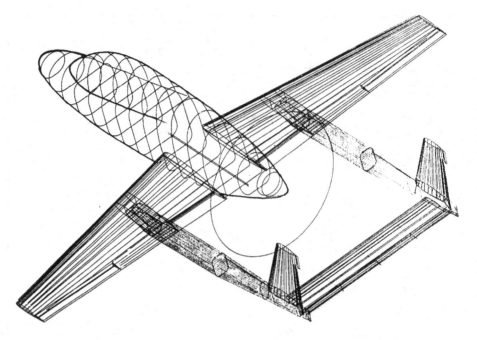

Fig. 9.9 T-84 turbo-prop training aircraft.

9.5.1.3 Modification

CAD systems provide many functions, such as delete or edit, that allow the user to modify the current display or design.

9.5.1.4 Draughting Aids

CAD system dimensioning capabilities conform to industry standards. The dimension function automatically generates dimensions, labels and associated lines related to the various entities of the part. Dimension types that can be generated include linear, angular, radial and diametrical, concentric circle, ordinate, and form and positional tolerance. The lettering/symbols functions are used to generate textural notes, labels, identification symbols, centre lines and cross-hatching. Standard section lining symbols are available to indicate common classes of material.

Figure 9.7 shows a typical drawing produced by a student as part of a space launcher group project and Fig. 9.8 shows a flap mechanism.

9.5.2 Three-Dimensional Modelling

Most serious CAD users graduate to 3-D models of components. These allow better visualization of parts and assemblies and are particularly useful for checking interferences and interfaces. The simplest 3-D models are of the wire frame type and include:

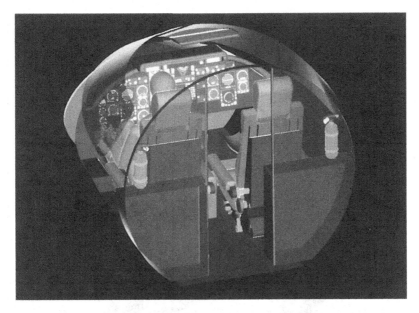

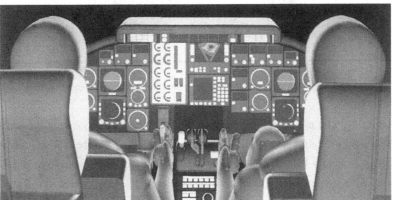

Fig. 9.10 E-92 project flight deck.

Ruled surfaces.

Surfaces of revolution.

Tabulated cylinders.

Fillet of blending surfaces.

Free-form 'sculptured' surfaces.

Fig. 9.9 shows a large model of a turbo-prop training aircraft. It was particularly useful in defining the complex wing/boom interface.

Wire-frame models are very efficient in terms of use of computer power. A modest machine can support the simultaneous use of four or five workstations; however, complex shapes can

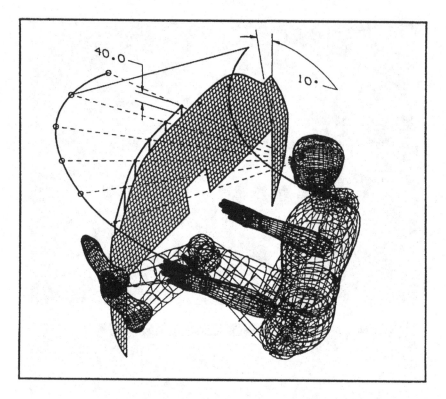

Fig. 9.11 Articulated model showing instrument panel and windscreen.

become confusing. The use of colour terminals can help, but confusion is still possible. Some systems have a feature called 'hidden-line removal', which helps, but the best solution is to use a solid-modeller program. This is inherently expensive in computer usage, but gives excellent visualization. Most systems have shading and alternate light sources and can show potential clashes between adjacent components.

Figure 9.10 shows a model of an executive aircraft forward fuselage. Different components have been constructed on different electronic layers in the system. It is thus easy to remove layers to get inside the aircraft. This process may be likened to over-laying a number of transparent sheets of paper, each with one CAD component, placed in the correct geographical position, relative to a set of darums.

Figure 9.11 shows one of Cranfield's articulated human figure models, that may be scaled to any size. The joints are capable of movement.

It is possible to enhance the model to build-in joint movement restrictions and strength capabilities. This has obvious application for modelling pilot forces and maintenance actions. Figure 9.12 shows the use of model to show large- and small-scale passengers in a cabin. Figure 9.13 shows a CAD model of the Cranfield LNG-14 airliner [65], which proposed the use

Fig. 9.12 E-92 cabin interior.

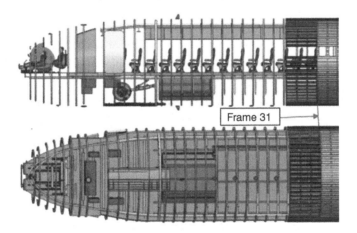

Fig. 9.13 LNG-14 forward fuselage.

of liquefied natural gas fuel to reduce environmental impact. The model was used to develop a methodology to perform zonal safety analyses at the preliminary design stage.

Several of the 3-D models have mechanism packages that show movement, and some of them are animated. These are particularly useful for the design of items with complex kinematics,

Fig. 9.14 Cranfield Demon CAD model.

such as landing gears. There is much work being performed to combine artificial intelligence with CAD systems in order to integrate design and manufacture. Figure 9.14 [44] shows a CAD model of the Cranfield Demon UAV, of which the author was chief engineer (see also Section 11.5.6). This was an interim model that was vital in the development of such a densely packed aircraft.

CHAPTER 10   THE SHAPE OF THINGS TO COME: SHOULD THE PROJECT CONTINUE?

10.1 INTRODUCTION

This book has attempted to give an introduction to most aspects of aircraft design, with emphasis on giving the reasons for the shapes of aircraft, and descriptions of their constituent parts. The preface clearly stated that it was *not* intended to produce a book about aircraft conceptual design, as many good texts are currently available. Some of these are referred to in the previous chapter, and they should be used, with help, to produce aircraft conceptual designs. Conceptual designs, however, are not usually end-products, but merely important steps in the whole design–manufacture–operation cycle.

Conceptual designs must be objectively assessed to see if they warrant the significantly increased expenditure that would be required to perform preliminary and detailed design. A vital precursor to this decision-making process is a clear definition of the characteristics of the conceptual design, and this is discussed in Section 10.2. Section 10.3 shows the options that are available following a conceptual design process and Section 10.4 describes simple decision-making techniques. The final section of this chapter gives an example of a conceptual-design definition, and describes the decisions that were made about its future.

10.2 CONCEPTUAL DESIGN DEFINITION

It is important to summarize clearly the results of the conceptual design process by providing the following types of information:

(i) Configuration description – This will be defined by means of conventional engineering drawings, tabular data, computer models, physical models or a combination of these methods, together with a brief description of the rationale of the design.

(ii) Mass estimates – These will include gross and empty masses, together with component mass estimates to be used as targets for subsequent work.

(iii) Performance estimates – Drag and lift predictions will be produced, together with powerplant data to give payload-range curves, field-performance, and direct operating cost targets for commercial aircraft. Military aircraft will also require predictions to be made of sortie performance, and point performance criteria such as attained turn rate, sustained turn rate and specific excess power.

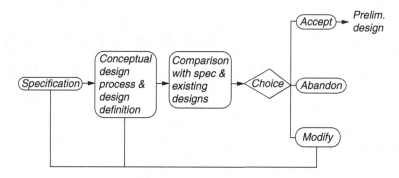

Fig. 10.1 Conceptual design decision-making process.

(iv) Reliability and maintainability targets – The direct operating cost targets, mentioned above, imply estimates of acquisition costs as well as reliability and maintainability (R and M) characteristics. These may be used to set R and M targets for subsequent design stages (Appendix C).

10.3 COMPARISON AND CHOICE

The next stage in the decision-making process is to compare the defined conceptual design with its specification, and with competing aircraft. This may often be a subjective process, but numerical values should be used as much as possible, to aid objectivity. This will be discussed below in Section 10.4.

The end of the decision-making process should follow from comparisons with the specification and the performance of competing aircraft, and will result in one of three possibilities, listed below and shown in Fig. 10.1:

(i) The design meets requirements and is sufficiently in advance of competitors to warrant further work and investment. The design should then proceed to the preliminary design phase and then to the detail design stage.

(ii) The aircraft does not meet the design requirement, or does not exceed competing aircrafts' performance, and has little chance of doing so after modification. It should be abandoned.

(iii) The design narrowly misses specification targets and should be modified and re-evaluated.

10.4 SIMPLE DECISION-MAKING TECHNIQUES

This is one of the most difficult tasks for any engineer and is often not amenable to numerical analysis. Several design characteristics will have been determined during the conceptual design process. These include initial studies of mass, performance and costs. These, alone, are insufficient to make a rational decision on the relative values of competing designs. Imponderables include aesthetics, technical risk, manufacturing capacity, etc. A designer needs some means of quantifying qualitative decisions. One suitable method is to list a series of design attributes and to give them weightings according to their relative importance. Individual designs may then be assessed and a mark given for each design, against each attribute. The attribute weightings and scores are then multiplied together to give a weighted score. The total attribute scores are added to give an overall rating, and thus the best design may be chosen.

Table 10.1 shows such an analysis to rank alternative designs of close air-support aircraft. It can be seen that two of the aircraft had very close results. Under such circumstances each design should be studied in more detail and the ranking process repeated on the basis of more accurate information.

Table 10.2 shows a decision-making chart for executive aircraft, taken from an advertisement [28], the data for which were themselves extracted from the *Business and Commercial Aviation 1993 Planning and Purchasing Handbook,* Conklin and de Decker Associates, Inc. The advertisement uses information on seven important business-jet parameters. In this example, the best aircraft in each attribute category was given the maximum rating of 10.0. The competing aircraft values were then scaled on a percentage basis, to give a relative rating.

For example, the Citation VII speed rating of 10.0 was factored by its speed of 470 mph and the Learjet 60's speed of 453 mph to give a rating of $10 \times (453)/470 = 9.64$ for the latter.

Inverse values apply when the acquisition and operating costs are rated. The overall ranking was the addition of all seven parameters and was termed the utility index.

10.5 EXAMPLE OF A CONCEPTUAL AIRCRAFT DESIGN DEFINITION DESCRIPTION: THE CRANFIELD A-90

10.5.1 Requirements for the Aircraft

The most important part of any design process is to get the requirements right! The first stage was to examine the main transport aircraft in use, or projected, together with suitable powerplants.

Several market surveys show that, despite short periods of recession, average annual growth rates for numbers of passengers will be 5.2% over the next 20 years. This growth is certain to strain airport capacity, which can only be alleviated by:

(i) Increasing aircraft load factors.
(ii) Increasing aircraft utilization.
(iii) Building airports and/or extra runways.
(iv) Improving air traffic control.
(v) Having larger aircraft.

Table 10.1 Close air support assessment summary

Attribute	Weighting	A Twin engine		B Single engine		C Low cost	
		Mark out of 10	Weighted mark	Mark out of 10	Weighted mark	Mark out of 10	Weighted mark
Overall cost	12	7.92	95	7.42	89	7.33	88
Low level performance	12	5.42	65	7.83	94	6.75	81
Detection probability	12	5.75	69	4.92	59	6.25	75
Battle damage tolerance	12	7.25	87	5.33	64	6.0	72
Payload/range	12	7.08	85	7.5	90	5.17	62
Maintainability, reliability/ repair	12	6.58	79	7.25	87	7.17	86
Field performance	7.5	7.07	53	8.8	66	8.93	67
Low project risk	15	8.0	120	7.0	105	6.2	93
Total			653		654		624

Table 10.2   Decision-making chart for business aircraft

Seating (typical)		Rating	Cabin size (cu ft)		Rating
1. Hawker 800	8	10.0	1. Hawker 800	604	10.0
2. Citation VII	7	8.8	2. Learjet 60	453	7.5
3. Learjet 60	7	8.8	3. Citation VII	438	7.3
4. Astra SP	6	7.5	4. Astra SP	368	6.1
Payload with max. fuel (lbs)		**Rating**	**Range (NBAA IFRR*)**		**Rating**
1. Hawker 800	1740	10.0	1. Astra SP	2668	10.0
2. Citation VII	1268	7.3	2. Hawker 800	2480	9.3
3. Learjet 60	1203	6.9	3. Learjet 60	2365	8.9
4. Astra SP	1060	6.1	4. Citation VII	1776	6.7
Speed (max)		**Rating**	**Operating cost (hourly)**		**Rating**
1. Citation VII	470	10.0	1. Learjet 60	928	10.0
2. Astra SP	463	9.9	2. Astra SP	947	9.8
3. Learjet 60	453	9.6	3. Hawker 800	1008	9.2
4. Hawker 800	442	9.4	4. Citation VII	1050	8.8
Acquisition cost ($M)		**Rating**	**Utility index/overall ranking**		**Rating**
1. Astra SP	7.54	10.0	1. Hawker 800		65.5
2. Learjet 60	8.30	9.1	2. Learjet 60		60.8
3. Citation VII	8.95	8.4	3. Astra SP		59.4
4. Hawker 800	9.95	7.6	4. Citation VII		57.3

* National Business Aircraft Association, Instrument Flight Rating Resources.

The design of a larger aircraft was the subject of the A-90 study. The market study [29] showed the need for a 500-passenger, short-haul airliner to fit between the capacities of current aircraft. The major features of the specification were:

  (i) 500 mixed class passengers.
 (ii) Carriage of underfloor standard LD3 containers and optional main deck cargo door.
(iii) Passenger and bag range of 2000 nmiles with FAR reserves, with maximum cruise speed greater than 340 knots
 (iv) All-up mass take-off from 8300 ft, and landing on 5650 ft, airfields at ISA, sea level conditions.
  (v) Runway loading less than that of the Airbus Industrie A330.

10.5.2 Configuration Description

The full conceptual design process for the A-90 is described by the author [29] and the parametric study is reproduced as Appendix B of this book. Figure 10.2 shows the three-view

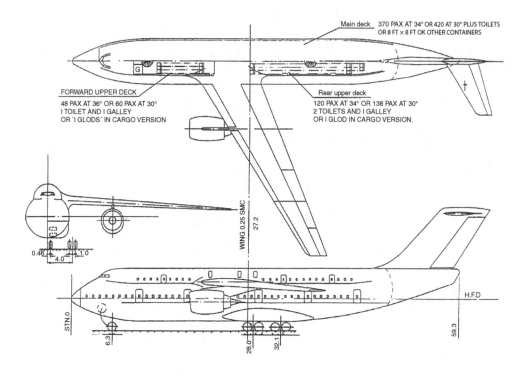

Fig. 10.2 A-90 general arrangement.

general arrangement drawing of the aircraft, whilst Table 10.3 gives the leading dimensions of the aircraft.

10.5.2.1 Wing

The modest quarter chord sweep of 30° combined with supercritical wing sections gave a maximum cruise of M 0.86. The wing aspect ratio is just less than 9 and the outer wing and centre-wing torsion box gave sufficient fuel volume for an optional range of 3500 nmiles, when combined with the trimming fin fuel tank. The wing utilizes variable camber leading- and trailing-edge flaps to improve

Table 10.3   A-90 leading dimensions and masses

Wing span	57 m (187 ft)
Wing span – folded	40 (131 ft)
Gross area	361.94 m^2 (3896 ft^2)
Quarter chord sweep	30.0°
Aspect ratio	8.98
Supercritical aerofoil t/c = 14%; root, 10.2%	MAC, 8% TIP
Anhedral	3.0°
Overall fuselage length	59.3 m (194.5 ft)
Maximum width	6.56 m (21.5 ft)
Height to tailplane	18 m (59 ft)
Passenger capacity (mixed)	500
Passenger capacity (all tourist)	620
Powerplant	RR Trent 800 series
All-up mass	211075 kg (466 923 lb)
Normal fuel mass	44 500 kg (98 018 lb)

cruise efficiency and to provide gust manoeuvre alleviation. They will provide scope for future development of increased-mass aircraft, as the camber may be used to generate a wide range of lift coefficients using the same wing size.

10.5.2.2 Fuselage
The cross-sections of current aircraft were examined and it was decided to use a 'double-bubble' fuselage, with a lower-lobe of similar width to that of the Boeing 747, with an upper lobe of similar width to the Airbus A320. It was felt that this would give a reasonable fuselage length/diameter ratio (which was later confirmed as being correct). Figure 10.2 includes a side view of the final fuselage configuration. All-economy seating has a capacity of 620 passengers. The main deck can accommodate two rows of 8 ft × 8 ft × 20 ft containers in a cargo version.

10.5.2.3 Powerplant Layout Choice
The first configuration choices were made during the parametric study (shown in Appendix B of this book). They were that twin turbo-fan engines would be used. This was considered to be the most cost-effective solution for a high-subsonic speed airliner.

The choice of wing-mounted podded engines was made for the following main reasons:

Engine position relatively close to the aircraft's fore-and-aft CG which allows payload and fuel loading flexibility.

Wing weight saving due to inertia bending relief from the engines, due to their mass partially counteracting the upward lift produced by the wings.

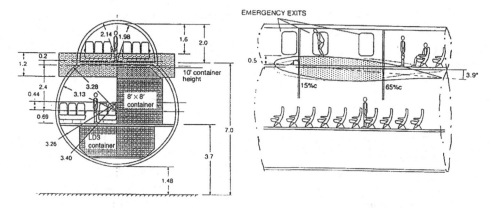

Fig. 10.3 Fuselage cross-section of A-90 aircraft.

Simple fuel feed from wing fuel tanks to the engines.

Relatively easy access for engine maintenance.

Avoidance of weight penalties and potential acoustic fatigue problems associated with fuselage-mounted engines.

The disadvantages of wing-mounted engines, relative to fuselage-mounted engines, were inferior noise, interference with wing high-lift systems, and problems of yaw control following engine failure. It was felt, however, that the advantages of wing engines outweighed their disadvantages. The chosen engines were taken from the Rolls-Royce Trent family.

10.5.2.4 Wing Position

The original choice was between high and low wing, because of the problem of providing wing bending carry-through structure in the fuselage. A mid/high wing position was chosen and Fig. 10.3 shows a computer-aided design model of the aircraft. The wing structure passes through the fuselage at the intersection of the upper and lower fuselage lobes. There is a clearance of more than 8 ft between the wing lower surface and main-deck floor for uninterrupted loading of 8 ft containers. The upper deck is partially divided by the upper-surface of the wing, but there is sufficient space for emergency transit between each half of the deck and access to the over-wing exits.

This arrangement thus combines the low interference drag between fuselage and wing of a mid-wing and the structural efficiency of a high-wing.

Other advantages of this layout are:

Good ground clearance for the under-wing pods of the 10 ft diameter engines. (This was the primary reason for this choice.)

The fuselage is closer to the ground because of less severe engine clearance problems. This eases cargo loading, passenger loading, and emergency passenger evacuation. The low

fuselage facilitates the conversion of the aircraft to a military transport role, with
loading ramps.

Easier access for fuselage maintenance, lavatory servicing and galley replenishment.

Main deck passenger vision is improved, except directly next to engines.

The disadvantages of this layout, relative to the more usual low-wing layout are:

The ditching properties are not as good. Ditching, or landing in water would be an extremely
rare event, and the upper-deck exits should alleviate or eliminate this problem.

The high-wing arrangement leads to fuselage mounted main landing gears. These are
installed in fuselage blisters, with added weight and drag. Such a big aircraft, even
with a low wing, however, would still require one or two fuselage-mounted main legs
and also require larger landing-gear fairings.

10.5.2.5 Tail Configuration

The mid/high wing location made a fuselage-mounted tailplane unattractive due to down-wash.
A high T-tail was chosen and mounted in such a position as to minimize deep-stall problems.
This arrangement led to the advantage of increasing the tailplane moment-arm, thus reducing
tailplane size. The tailplane also acts as an end-plate, and increases the fin's aerodynamic
effectiveness. The disadvantages of this choice are those of increased fin and rear fuselage loads
due to asymmetric tailplane lift, and maintenance access to a fin some 18 m (60 ft) above the
ground!

The latter problem was reduced by specifying an internal maintenance ladder forward of
the fin front spar, where the width was some 0.8 m (2 ft 6 ins) at the intersection with the tailplane.

10.5.2.6 Landing Gear

The nose unit has a conventional twin-wheel forward-retracting configuration. There are four
fuselage pod-mounted main units, each with a four-wheel bogie.

10.5.2.7 Systems

Conventional avionic systems are used, but the 'active' variable camber flaps require a fast-acting
flight control system. Fibre-optic signalling was chosen. The large passenger capacity required
most of the engine bleed air to feed the environmental control system. Electro-impulse de-icing
was used and most secondary power was produced electrically.

10.5.3 Performance Estimates

Drag estimates were made using simple component build-up methods and ESDU data sheets aided
the estimation of lift, and stability and control derivatives.

10.5.3.1 Lift Characteristics

Maximum lift coefficient:

Basic wing 1.20

+ variable-camber devices (take-off) 1.8

+ variable-camber (landing) 2.5

10.5.3.2 Drag Characteristics

Drag polar:

Cruise condition at M 0.86 at 12 000 m altitude

$C_D = 0.02 + 0.047C_L^2$

Take-off at sea level, undercarriage and flaps extended

$C_D = 0.0488 + 0.047C_L^2 + 0.053\Delta C_L^2$

Landing at sea level, undercarriage and flaps extended

$C_D = 0.0948 + 0.047C_L^2 + 0.053\Delta C_L^2$

Where ΔC_L is increment in C_L due to flap deflection

10.5.3.3 Pitching Moment Characteristics (Low Speed)

Pitching moment coefficient at zero lift:

Wing alone, C_{Mo} –0.06

Increment due to body C_{M_B} –0.015

Pitching moment increment due to flaps:

Take-off setting, $\Delta C_{M_{FTO}}$ –0.09

Landing setting, $\Delta C_{M_{FL}}$ –0.17

Location of overall wing-body aero centre:

Percentage of wing standard mean chord (SMC), whole aircraft 20.3%

Includes:

Forward shift due to basic fuselage 5.6%

Forward shift due to nacelles 7.0%

10.5.4 Mass Estimates

A semi-empirical mass estimating method was used to predict the overall mass of the aircraft. It was subdivided to system-level to give the targets shown in Table 10.4. These figures may be compared with the more exact estimated masses also shown in the table. The latter were obtained after the completion of preliminary and some detail design.

Table 10.4   A-90 component mass estimates and targets

Component	Estimated mass (kg)	Target mass (kg)
Wing group-structure (inc. actuators)	20 790	17 920
Fuselage structure	28 315	31 554
Fin and rudder (inc. actuators)	1 934	1 200
Tailplane and elevators (inc. actuators)	1 180	1 020
Undercarriage	8 420	7 855
Pylons	2 160	21 00
Structure	62 799	61 649
Engines, powerplant structure and accessories	14 008	15 509
Fuel system	1 324	1 215
Flying control system	1 395	1 551
Hydraulics	32	1 411
Electrical system	1 680	3 350
Auxiliary power unit (APU)	260	230
Instruments and avionics	1 100	1 120
De-ice	397	830
Fire protection	80	620
Furnishings	7 100	7 100
Environmental control system	1 423	2 078
Paint	180	180
Systems and equipment	14 971	19 685
Manufacturers equipped mass	91 778	96 843
MEM tolerance	7 002	1 937
Crew and provisions	3 690	3 690
Seats, emergency equipment pax service	11 800	11 800
Nominal operating empty mass (OEM)	114 270	114 270
2% mid-life mass growth allowance	2 285	2 285
Pallets and containers	2 520	2 520
Operating empty mass	119 075	119 075
500 passengers and baggage	47 500	47 500
Fuel at above payload	44 500	44 500
Maximum all up mass	211 075	211 075

10.5.5 Payload-Range Characteristics

The drag estimates, engine data and other information were used to derive the payload-range diagram, as shown in Fig. 10.4.

The A-90 is the original aircraft, described above. The A-90A and B are variants, redesigned with different technology standards [30].

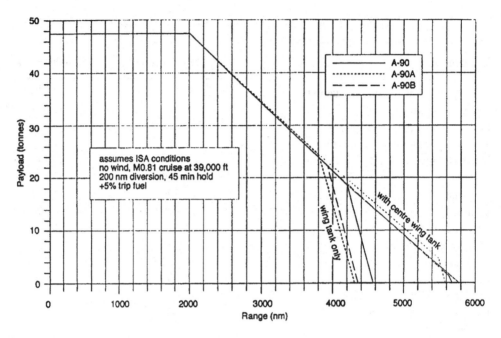

Fig. 10.4 A-90 payload-range diagrams.

10.5.6 Cost Estimates

The cost prediction methods of Roskam [31] were used to estimate the acquisition and direct operating costs of the aircraft. There was considerable doubt about the development cost effects of the advanced technology used on the A-90. This was investigated by Lim [32]. The acquisition costs for the original A-90 were US$90 million, based on a production run of 500 aircraft. The A-90A was similar to the original aircraft, but used conventional flaps and a re-sized wing. The A-90B was the same as the A-90A, but used more conventional materials and technology. The predicted acquisition costs per aircraft were US$84 million for 500 A-90A aircraft and US$73 million for 500 A-90Bs. Figure 10.5 shows relative direct operating costs of the three models and those of real aircraft, using consistent methods.

10.5.7 Reliability and Maintainability Targets

The dispatch reliability method described in Appendix C was used to set targets for dispatch reliability. These are shown in Table 10.5 below. A prediction method for civil aircraft maintainability was not available at the time of the A-90 study. Dispatch reliability is defined as unity-delay rate, where:

$$\text{Delay} \quad \text{rate} = \frac{\text{No. of delay} > 15\,\text{min} + \text{cancellations}}{100\,\text{departures}}$$

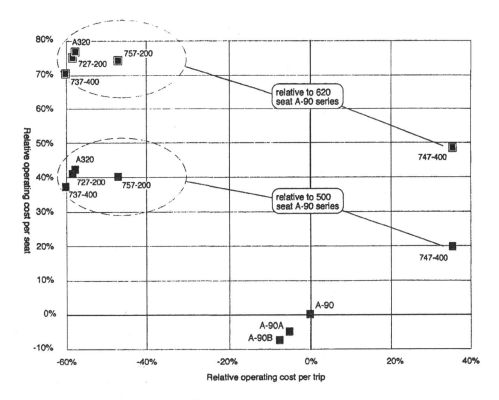

Fig. 10.5 A-90 operating cost comparisons.

These delays are those caused by technical problems, but non-technical delays are excluded, i.e. those due to air-traffic control or late passengers. Civil aircraft systems are allocated what are termed ATA chapter codes, where ATA stands for Air Transport Association of America. The method in Appendix C was used to predict targets for delays caused by defective equipment from each aircraft system. For example, the landing-gear system should not cause more than 0.41 technical delays per 100 departures. The overall aircraft dispatch reliability target was predicted to be:

$$1 - \frac{2.87}{100 \text{ Departures}} = 0.9713$$

therefore the dispatch reliability target for the whole aircraft is 97.13%

10.6 PROGRESS OF THE A-90 PROJECT BEYOND THE CONCEPTUAL DESIGN STAGE

The conceptual design was performed by the author, and showed considerable promise, relative to existing aircraft, but it had several areas of moderate technical risk. It was decided to

Table 10.5   Delay rate targets for individual systems

ATA Chapter No.	Description	Delay rate per 100 departures
21	Air conditioning	0.12
22	Auto, flight	0.03
23	Communications	0.03
24	Electrical power	0.10
25	Furnishings	0.06
26	Fire protection	0.05
27	Flying controls	0.32
28	Fuel system	0.10
29	Hydraulic power	0.18
30	Ice protection	0.01
31	Instruments	0.02
32	Landing gear	0.41
33	Lights	0.07
34	Navigation	0.15
35	Oxygen	0.01
38	Water/waste	0.02
49	APU	0.09
52–57	Structures	0.24
71–80	Powerplant systems	0.86
Total		2.87

proceed with preliminary and early detailed-design study by means of the Cranfield 1990–91 MSc course in aerospace vehicle design. Twenty-three students and five members of staff performed an eight-month group design project and obtained considerable confirmation of the design [30].

The author presented the design at several conferences and caused some controversy by the choice of the wing position.

The A-90 formed the basis of the conceptual design of the A-94 project. The fuselage section and nose fuselage were retained, but a new wing was designed for the new long-distance range requirements for the stretched aircraft. Extra, parallel, fuselage sections were designed to accommodate an increase in passenger capacity up to 600 mixed-class or some 750 in high-density. This project also formed the basis of a Cranfield group design project in 1994–95, studied by 28 students and six members of staff. The results are described [33] and a CAD model is shown in Fig. 10.6. The results were promising, but highlighted many areas requiring further investigation,

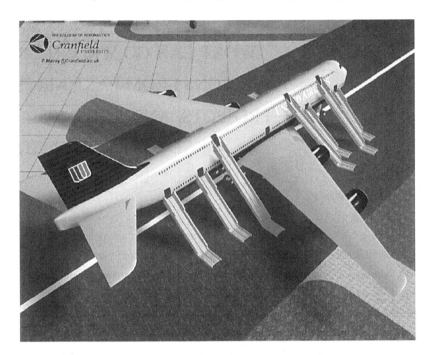

Fig. 10.6 A-94 project with escape slides deployed.

particularly in terms of emergency evacuation, airport compatibility, landing gears and manufacturing requirements.

It would be nice to take the A-94 to production, but the development cost of US $10–14 billion is beyond the budget of most universities! Perhaps some of the ideas might be incorporated by some aircraft manufacturers.

CHAPTER 11

WHAT CAN GO WRONG?: SOME LESSONS FROM PAST AIRCRAFT PROJECTS, AND A GLIMPSE INTO THE FUTURE

11.1 INTRODUCTION

The design, manufacture and operation of aircraft are very complex and potentially risky activities. There is a natural tendency towards trusting tried and tested configurations, and adopting an evolutionary approach. There are many cases, however, of revolutionary design changes, such as the advent of the jet engine, or the delta wing. Revolution involves more risk, but can be very rewarding. The history of aircraft design has had many examples of excitingly innovative, not to say weird designs. A minority of these are successful, some marginally successful, but many have been failures, some of them disastrously expensive failures! This chapter will describe a number of different projects and attempt to classify the reasons that led to their difficulties.

On a more positive note, the chapter will conclude by describing a number of innovative designs that might have a great influence on the future of aeronautics – positively or negatively!

11.2 AIRCRAFT THAT SUFFERED FROM REQUIREMENTS THAT WERE TOO RESTRICTIVE, TOO AMBITIOUS OR WERE CHANGED DURING DEVELOPMENT

Ward [34] contains many examples of such projects from the depressing record of British aircraft developments since 1945, and several examples will be repeated here, both civil and military. Other sources were used for later aircraft.

11.2.1 The Hawker Siddeley Trident

This aircraft (Fig. 11.1) had many years of successful service in the UK and in China, and was a pioneer in the field of automatic landing, but sales were disappointing. British European Airways (BEA) were initially offered a short-haul jet transport, (the DH121) which would have carried 110 passengers over stage lengths of up to 1800 nmiles. The manufacturers' projections in 1958 predicted sales of 550 aircraft. According to Ward [34] BEA had been examining market trends that showed a marked slip in 1958–9 and thus considered that the DH121 was too big. They required that the aircraft only have 97 seats over an 800 nmile range. The Trident was designed round this specification, as was the Spey turbo-fan engine. The aircraft satisfied the requirement,

Fig. 11.1 The Hawker Siddeley Trident.

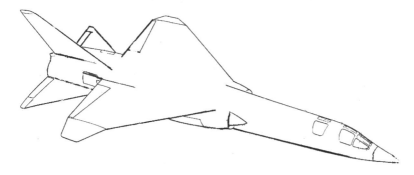

Fig. 11.2 The BAC TSR-2 bomber.

but it and the engine had limited development potential. The Boeing 727 design team developed a competing tri-jet to almost the same requirements as for the original DH121. Both aircraft were developed into several models, but the Trident sold 117 aircraft as against the Boeing 727's 1831!

11.2.2 The BAC TSR-2 Bomber

It is difficult to separate the technical, financial and political factors that had significant impacts on military aircraft design in Britain in the 1950s and 1960s. In 1957, the Conservative Defence Minister, Mr Duncan Sandys, decreed that the vast majority of military aviation tasks would be performed by guided missiles. A very large number of promising projects were cancelled immediately. The industry was badly affected, but managed to recover to produce several design projects by the early 1960s, one of which was the TSR-2 bomber (Fig. 11.2). This project was cancelled on the grounds of cost by the Labour Defence Minister, Mr Denis Healey. This was in the same period

Fig. 11.3 The TSR-2 at Cranfield.

as the cancellation of the AW 681 V/STOL military transport and the P1 154 supersonic derivative of the Harrier V/STOL aircraft.

The cancellation of the TSR2 was a complex process which is described in Ward [34] and also in more detail in Barnett-Jones [35]. Both authors acknowledge that the TSR-2 was the most challenging aerospace project to date. The OR 343 requirement called for:-

(i) Speeds of M 1.1 at 60 m altitude and M 2+ at medium altitudes.
(ii) A radius of action of 1000 nmiles, including 100 nmiles supersonic at altitude and 200 nmiles out and return at M 0.9 at sea level.
(iii) Ferry range greater than 2800 nmiles.
(iv) Take-off ground roll of not more than 550 m from rough strips.
(v) Fully automatic, accurate, navigation system, radar, infra-red and photographic systems.

Further complications came from the then new systems approach to weapons-system design and joint design by previously competing companies. There was much involvement by government committees and the aircraft became ever more complex and expensive. Prototypes were built and successfully flown, but costs escalated and inter-service rivalry clouded the issue, as both the Royal Air Force and the Royal Navy had conflicting claims on a very limited defence budget.

The TSR-2 was cancelled and all production aircraft were destroyed in the factory or on artillery ranges. A few prototypes were preserved and placed in museums, and at the College of Aeronautics at Cranfield (Fig. 11.3). That aircraft was subsequently transferred to the Imperial War Museum at Duxford, Cambridge, England.

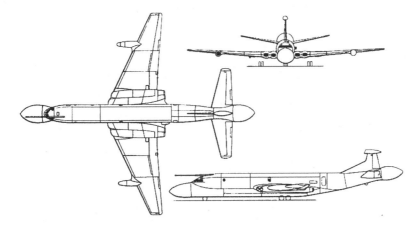

Fig. 11.4 The BAE SYSTEMS AEW Nimrod.

The role of the TSR-2 was to have been taken over by the American F–III programme, but this also was cancelled and the RAF had to wait a long time for the multinational Tornado aircraft. That aircraft, however, benefitted much from the expertise and technology developed on the TSR-2 programme, as did the Concorde supersonic airliner.

11.2.3 The Airborne Early Warning Nimrod

The principles of airborne early warning (AEW) have been mentioned in Chapter 7. The US Air Force and NATO have met the AEW need by mounting a large rotating radar scanner on the top of a converted Boeing 707 airliner (AWACS). This has strengths and weaknesses, in that it is relatively simple, but the airframe masks certain areas of the radar-system's field-of-view.

GEC/Marconi and BAE SYSTEMS chose a different approach in the design of the AEW Nimrod (Fig. 11.4).

The aircraft would have used a pair of nose and tail-mounted radar antennaes, each having a 180° field of view without obstruction. The author performed some early layout work for this project, based on a conversion of the Nimrod Maritime reconnaissance aircraft which was, itself, a development of the Comet airliner. This arrangement gave a splendid field of view, but the sophisticated avionic systems led to the need for a more powerful generating system. Larger generators and avionic systems led to the need for a much more capable environmental control system, to cope with the increased heat. This process led to the classic weight-growth snowball effect. It was reported that the problem was compounded by customer requirements – changes and unsuitable radar-system performance. All of those factors led to cost growth and cancellation, and the ordering of the competing design, the Boeing AWACS, in the late 1980s.

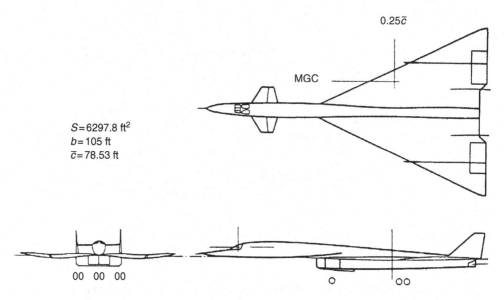

$S = 6297.8 \text{ ft}^2$
$b = 105 \text{ ft}$
$\bar{c} = 78.53 \text{ ft}$

Fig. 11.5 Rockwell XB-70.

11.2.4 The Rockwell XB-70, Boeing B2707 and NASP

The challenging demands, weight growth and cost escalations were not confined to British aircraft. It was a world-wide phenomenon, and can be a consequence of unrealistic requirements and high expectations of future technology. The Rockwell XB-70 was a huge strategic bomber, designed to fly at M 3.0. The kinetic heating effects are considerable at this speed, so conventional aluminium alloys cannot be used and alternatives had to be developed. This increased risk and cost. The XB-70 entered the cost–weight spiral and never went further than the prototype stage (Fig. 11.5).

Mach 3.0 was also the chosen cruise regime for the American competitors for the supersonic transport market. Many aircraft were designed as projects during the late 1960s and early 1970s. They were to fly fast and farther, with more passengers than the Anglo-French Concorde (see later).

Many hundreds of millions of dollars were spent on these projects, culminating in the Boeing B2707. This was a formidibly challenging project that followed the weight–complexity–cost growth route, and never reached the prototype stage.

This path was repeated in the USA in the 1980s on a project variously named 'The Orient Express' or the 'National Aerospace Plane (NASP)'.

These were various projects aimed to fly some 300 passengers on trans-Pacific routes at Mach numbers greater than five. These were extremely challenging aims and although some technology demonstration has been done, there is little yet to show for all the effort expended.

Fig. 11.6 The BAE SYSTEMS/Aerospatiale Concorde.

11.3 Projects that were Overtaken by Events

Large aircraft projects are complex and often have a long gestation period. In a number of cases world events have changed the circumstances that were important at the time of the project's inception.

11.3.1 BAE SYSTEMS/Aerospatiale Concorde

The Concorde (Fig. 11.6) has been an outstanding technical success and the small number of aircraft in service have accumulated significantly more supersonic flying hours than all the rest of the aircraft in the world. It operated reliably, primarily on the North Atlantic routes, but it has not been the commercial success that was initially hoped for.

As the first supersonic transport, it had to push back many frontiers in aerodynamics, propulsion, systems, structures and operations. A careful collaboration programme lasted for many years in the 1950s and 1960s. There were effectively three different types of aircraft in terms of prototypes, development aircraft and production models. Environmental science and the environmental lobby were in their infancy during the early design phases, but they developed rapidly. It had been assumed that the aircraft could operate supersonically throughout the world, but environmental pressure to limit the effects of sonic boom led to the restriction of supersonic flight to over-water sectors. This negatively affected operational efficiency and excluded the aircraft from a number of overland routes, thus reducing its potential market. Concorde's initial take-off and landing noise were considered to be too loud and much effort went into reducing engine noise, but Concorde's noise attributes would be unacceptable for future supersonic transports. Fuel prices increased rapidly during Concorde's development, further reducing its economic viability, relative to early expectations. All of these factors led to the operation of only 14 in-service aircraft split

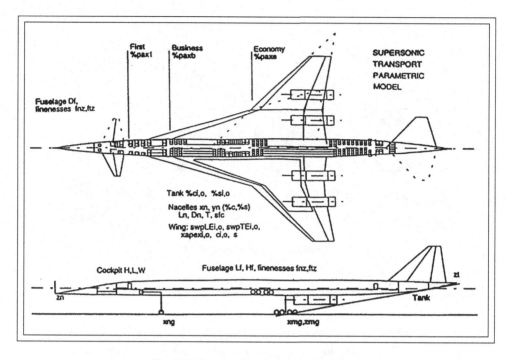

Fig. 11.7 Supersonic transport configuration.

between British Airways and Air France. They were, however, flagship aircraft flying at premium fares and gave great prestige to their passengers and operators.

Super-Concordes are in the early design stages on both sides of the Atlantic and current schemes are for aircraft flying at speeds similar to Concorde's M 2.0, but with trans-Pacific rather than trans-Atlantic ranges. They will carry some 300 passengers, rather than Concorde's 108, to reduce seat-mile costs. Variable-cycle engines will be required so that they can be efficient in cruise, whilst meeting stringent take-off and landing noise requirements. Figure 11.7 shows a typical project of the late 1990s.

11.3.2 The Northrop F-20 Tigershark

The F-20 (Fig. 11.8) was an attempt to develop a cheap, effective lightweight fighter as a replacement for the very successful Northrop F-5 fighter. The market was the countries that already operated the F-5, particularly in South-east Asia. The F-20 was a very significant modification to the F-5, starting with the same airframe, but substituting a single modern F404 for a pair of older turbo-jets. The nose of the aircraft was reconfigured to improve agility, and a totally new avionics and flight control suite was developed. The result was a very capable fighter. At the start of the F-20 development, the US government had imposed an export embargo on the sale of its most

Fig. 11.8 The Northrop F-20 Tigershark.

modern fighters, such as the F-16. This included many of the target countries for the F–20. Wilson [36] listed some of the reasons why the project did not succeed. The aircraft matured too late and by this time the embargo on the F-16 had been lifted for some of the potential F-20 customer countries. There was no specific home market for the F-20 in the USA, and potential customers naturally asked 'why hasn't your country bought it?'. The project was a private venture project with no strong financial committment from potential customers.

Prototypes were built and flown, but no production aircraft were sold, and the project stopped.

11.3.3 The Saunders-Roe Princess Flying Boat

This large and stylish flying boat (Fig. 11.9) was designed, built and tested in the late 1940s as an updated continuation of the pre-WWII passenger flying boats. Unfortunately for flying boats, the world was a different place after the war! Many large airfields had been built for bombers and became available to the new, more efficient land planes. Flying boats do not need runways, but they do need mooring, and passenger loading and maintenance facilities, which are expensive for large aircraft. Flying boats have to carry the mass penalty of a reinforced hull, which has to resist the severe structural loads caused by operations from water. The Princess also used what is termed a 'step' in the hull which is used to reduce hydrodynamic suction on take-off. It does, however, also lead to aerodynamic drag penalties with consequent mass, fuel and cost penalties.

All of these factors, combined with a complex propulsion system, meant that the Princess could not compete economically with the land planes of its generation, so the project was cancelled in 1954. There is, however a niche market for smaller flying boats and float planes, where it is uneconomic to develop airports. Chicken [37] describes some recent work that has developed modern project-design methods for new amphibious aircraft.

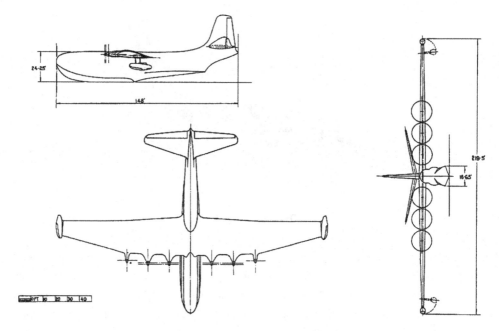

Fig. 11.9 The Saunders-Roe Princess.

11.4 A Step Too Far or Too Soon?

There are many examples of novel concepts that seem to offer dramatic performance improvements that are unrealized. All true research involves risk and often reveals unexpected snags. These can lead to complications or show the need for technology that is not yet available, or becomes too expensive to acquire. A few of such projects will be described here.

11.4.1 Forward Sweep Wing: The Grumman X-29

Transonic aerodynamic performance can be improved by a number of means, including wing sweep-back or sweep-forward. The former is commonplace, but the latter has been restricted by the phenomenon of wing divergence. A flexible forward-swept wing will twist nose-up, relative to the root section, as it is subjected to aerodynamic lift loads. The increased twist increases the local angle of attack, thus further increasing the lift load, until failure. It is possible to combat this effect in a conventional light-alloy wing by increasing the wing's torsional stiffness by using thicker material. This, however, is usually prohibitive in terms of mass. The advent of carbon-fibre composite materials suggested a possible solution for forward-swept wings. The designer may control the orientation of the directions of several layers of the carbon fibres, such that the wing will twist nose-down when subjected to wing lift loads. This concept was incorporated into the experimental Grumman X-29 aircraft (Fig. 11.10).

Fig. 11.10 The Grumman X-29.

Two prototypes were built and flight-tested, but as yet, no production aircraft have been developed. The author performed a detailed design of an aircraft with a similar wing layout and then had it subsequently studied as the S-83 group-design project [38]. The study showed that the aircraft should perform very well in the transonic regime, but that the wing produced too much drag in the supersonic region. The project was further complicated by the requirement to have VTOL capability, which led to an extremely complex aircraft (Fig. 11.11). It was felt that although the aircraft could have been built, it would not have been cost-effective.

11.4.2 The Hermes Spaceplane

The European Space Agency (ESA) decided to develop a manned space shuttle for exploration, and missions to support the operation of the proposed low-Earth-orbit space station. It was to have been vertically launched on the top of the Ariane V rocket. The Ariane V was simultaneously being developed for a number of other launch missions and had a fixed payload launch capability. This set a maximum limit on the launch mass of the Hermes that could not be exceeded (Fig. 11.12).

The launch mass, payload and crew requirements led to the specification of very sophisticated airframe and heat-shield materials, which required considerable development. As in many projects, more detailed design revealed the need for mass increases, at the expense of payload, because of the fixed launch mass. The issue was further complicated by increased complexity and mass associated with improved crew safety measures following the Space-Shuttle Challenger tragedy. These and other factors, led to the cancellation of the project in the early 1990s.

Fig. 11.11 The Cranfield S-83 project.

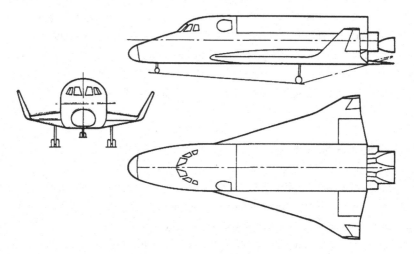

Fig. 11.12 The Hermes.

11.4.3 The Beech Starship

This aircraft (Fig. 5.1) was a radical aircraft designed to bring executive-jet performance with turbo-prop propulsion and economy. The bold design had an all-composite structure, a canard foreplane, twin pusher propellers and wing-tip fins. The construction was extremely innovative and led to pioneering techniques. It had the problems of being first in the field and had to overcome many certification hurdles, because of that. Some 50 aircraft were built, but sales were disappointing and production was stopped in the late 1990s.

11.5 Some Challenging Future Projects

University aeronautical departments have the luxury of being able to think laterally without having to satisfy share-holders, as do recently retired design professionals. The projects described here are some that were being examined in association with Cranfield University since the late 1990s, and might lead to new innovative designs.

11.5.1 Subsonic Airliners

Chapter 10 describes the relatively conventional A-90 500-seat short-haul and A-94 600-seat long-haul airliners. A number of alternative configurations have been studied, including blended wing body (BWB) designs.

11.5.1.1 Blended Wing Body Designs
The Cranfield BW-11 blended wing project, shown in Figure 11.13 [39]. The wingspan is constrained by the airport 80 m 'box', thus limiting its aspect ratio. Such configurations have significantly lower surface areas than conventional configurations, and offer large aerodynamic and payload flexibility benefits. The passenger cabins need to be pressurized, which leads to challenges in fitting cylindrical pressure vessels inside a wing profile. These can be solved by use of a series of joined cylinders, covered by an aerodynamic skin, or stiffened monolithic skins. Emergency evacuation issues and lack of passenger windows have been overcome by careful research into aisles and exits, and 'electronic' windows.

Figure 11.14 shows an image of the interior of the Cranfield BW-02, 250 seat long-range aircraft [40]. The emergency evacuation routes are through wing leading-edge doors, and from the rear of the cabin, to the under-surface of the aircraft. It uses two internal gas-generating powerplants, driving four fans on the upper surface.

Such aircraft configurations pose significant technical risks, so much research is needed. One of the most powerful ways of providing this is to develop subscale flying demonstrator aircraft. The Boeing/NASA/ Cranfield Aerospace Ltd. X-48B fling aircraft are prime examples of the success of such an approach. An earlier example is shown in Figure 11.15. This is the Cranfield University Kestrel BWB demonstrator, developed during the part-time masters course in aircraft engineering. Its take-off mass is about 130 kg, with a span of some 7 m, and it is powered by two AMT Olympus gas turbines. It was completed in conjunction with BAE SYSTEMS, and successfully flew in 2003. BWB aircraft have great potential, and they are being studied in many parts of the world.

Fig. 11.13 Cranfield BW-12 Blended Wing Body Project

11.5.1.2 Other Novel Concepts

There has, in recent years, been considerable emphasis on designing aircraft for reduced environmental impact. Reference 41 shows a few of the concepts investigated by Cranfield University. Aircraft designed for very long ranges need to carry enormous amounts of fuel, leading to a divergent spiral of mass, wing size, powerplant thrust, etc. Figure 11.16 shows an alternative concept that uses aerial re-fuelling of airliners. The receiving aircraft is a conventional airliner of the Boeing 787/Airbus A350 class, but optimized for a range of only 4000 miles. It may be refuelled to fly either 8000 or 12000 miles. The tanker aircraft would operate from 8–10 bases worldwide to refuel the airliners. The tanker approaches from below and behind, for efficient fuel transfer. Net fuel burn reductions are some 30% of 2000 technology airliners (including tanker fuel burn), but airworthiness issues must be resolved.

Figure 11.17 shows a concept of a 'box wing'. Its configuration has reduced span, good aerodynamic efficiency, wing structural benefits, but complex stability and control requirements. It should save about 40% fuel burn, relative to 2000 technology aircraft.

Aircraft skin friction drag may be lowered by ensuring that the airflow remains in the laminar region, rather than turbulent. This is very difficult to achieve when using swept wings to

Fig. 11.14 The Cranfield BW-02 BWB airliner.

Fig. 11.15 The Cranfield Kestrel BWB demonstrator on its first flight.

Fig. 11.16 The Cranfield MRT-7 airliner receives fuel from a tanker (below and aft).

Fig. 11.17 The Cranfield A-9 box-wing airliner.

delay compressibility in transonic flow. Swept wings therefore usually require suction to the wing leading edges, in what are termed hybrid laminar flow aerofoils. This is not a new concept, as a Lancaster bomber was flight-tested by the College of Aeronautics at Cranfield in the 1960s. It was fitted with a Handley-Page-designed small hybrid laminar flow vertical wing in the dorsal turret position. The Lancaster was subsequently passed to the Royal Air Force Battle of Britain Memorial Flight, and is still flying today! Hybrid laminar flow is currently being evaluated on both sides of the Atlantic. A simpler method of achieving laminar flow is to use a specially designed natural laminar flow aerofoil, as in the Cranfield A-6 Greenliner project, shown in Figure 11.18. This aircraft uses an unswept wing, so that it does not need expensive or heavy hybrid laminar flow equipment. This wing, however results in a cruise Mach number of 0.74, rather than the more usual long-range values between 0.83 and 0.85. The cabin is made more comfortable, but it remains to be

Fig. 11.18 Cranfield A-6 Greenliner.

seen if passengers will accept slower flight as part of the trade-off against cost and environmental pollution savings of some 40% relative to year 2000 technology aircraft.

11.5.2 Very Large Cargo Aircraft

Passenger travel was predicted to increase at a compound rate of 5% per annum in between 1990 and 2010. Cargo was predicted to grow at 8% per annum over the same period. A number of configurations have been proposed including the twin-fuselage aircraft shown in Fig.11.19, developed from the work by Roeder [42]. The total payload was to be 400 t disposed in the fuselages, each of which has a cross-section similar to the Airbus Beluga, a large cargo aircraft.

The project was to proceed from an initial single-fuselage aircraft with a payload of some 150 t. The aircraft in the figure would be developed using two of the fuselages, with stretch, and joined by the centre wing. The fuselages are some way from the aircraft centre line and given considerable wing bending moment reduction, with considerable weight saving. The landing gear and runway widths required by such an aircraft require some work on the airport infrastructure. It was envisaged that a relatively small number of ex-military airfields could be suitably modified at relatively modest cost.

Further work was carried out to investigate the feasibility of using the broad delta concept as a large cargo aircraft. This concept has considerable merit, as the cargo can be put in the wing, which has relatively smaller span than the twin fuselage, thus easing airport compatibility questions.

11.5.3 Wing in Ground Effect (WIGE)

It has been known for a long time that aircraft flying very close to the ground can operate at a higher aerodynamic efficiency than normal flights because of the ground effect. Many such designs have been proposed, particularly more recently in the Former Soviet Union.

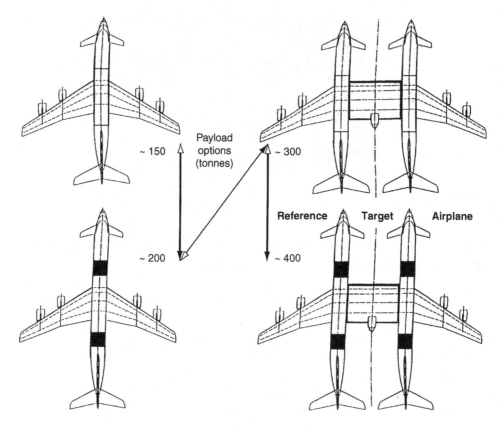

Fig. 11.19 Twin fuselage cargo aircraft.

Such craft are hybrids between aircraft and ships in that they operate at aircraft speeds over smooth surfaces – usually the sea. They are different from hovercraft, in that the lift on the craft is generated by aerodynamic forces on wings, not from an air-cushion provided by engines. The craft has very significant speed advantages over ships, but must have means of overcoming obstacles, such as very large waves. It therefore needs to be able to fly as an aircraft out of ground effect to over-fly obstacles. This makes it more suitable for overland flight than hovercraft, but does add complications. One such a project is shown in Fig. 11.20 [43]. It has a number of technology issues that must be solved, but it has great potential.

11.5.4 Supersonic Short Take-off, Vertical Landing Fighter

There has been considerable interest in the development of a supersonic fighter that also has the vertical take-off and landing attributes of the subsonic Harrier aircraft. The Cranfield S-95 project attempted to do this, with the additional requirement that the aircraft should have low radar and infra-red signatures, combined with the ability to operate from aircraft carriers. This was clearly a

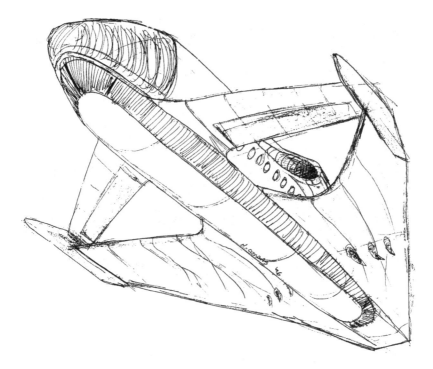

Fig. 11.20 Possible wing in ground effect aircraft.

challenging task. Conceptual, preliminary and detail designs were performed by means of individual study and group design project work. The resulting aircraft appeared to be feasible and should meet all of the performance requirements. It had a very large engine, posing considerable problems with removal for maintenance, which were solved.

Figure 11.21 shows the conceptual design computer model of the aircraft (top left). The centre-right image shows the preliminary structural and systems design of the tail-boom. The lower-right image is a CAD solid model of a fin pick-up frame, whilst the remaining image is of a structural finite-element model of the frame.

This figure therefore shows elements of conceptual, preliminary and detail design stages, and structural analysis.

The aircraft was promising, but was predicted to cost some US$80 million each in 1996. Cost-effectiveness studies would have to be performed to see if such a project should be produced.

11.5.5 Supersonic Business Jet

The Concorde has been the only successful supersonic transport aircraft, but had to be retired following a worldwide recession. The operational costs of such an advanced aircraft were deemed to be uneconomic.

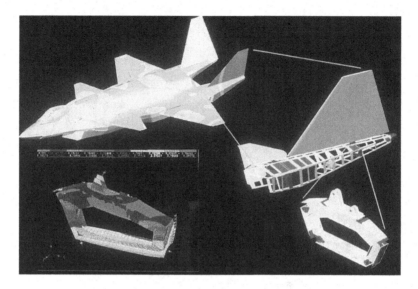

Fig. 11.21 The Cranfield S-95 project.

There have been numerous attempts to design new supersonic transports, often with a seating capacity of 300, with trans-Pacific range at speeds of M 2.0. The current business case for such aircraft has not been attractive.

There is a much stronger case for supersonic business jets, such as the Cranfield E-5 Neutrino, shown in Fig. 11.22. A sufficient number of passengers are prepared to pay premium prices for the time, saving, flexibility, prestige and security afforded by such an aircraft. The slender, small aircraft should be able to fly overland at M 1.2 and oversea at M 1.8 without problems caused by sonic booms. Significant engine design improvements are needed, however, to meet airport noise requirements.

11.5.6 The Cranfield Demon UAV Research Flying Demonstrator Aircraft

Chapter 4 of this book shows some of the roles being performed by, or planned for, UAVs. As with all aircraft, progress requires research and realistic demonstration before it can be risked on new aircraft projects. This was recognized by the UK ERSRC and BAE SYSTEMS when they sponsored the Flaviir multi-university research programme, led by Cranfield University [44]. Many technologies were successfully developed, but a major programme requirement was to demonstrate them on a realistic flying vehicle. The Demon aircraft (for which the author was chief engineer) was aerodynamically scaled up from the earlier Cranfield Eclipse aircraft, to the configuration shown as an early CAD model in Fig. 11.23.

The aircraft was designed and constructed at Cranfield University, with significant inputs from partner universities, and apprentices and other employees from BAE SYSTEMS. Fig. 11.24

Fig. 11.22 Cranfield E-5 Neutrino supersonic business jet.

shows the aircraft in a flight, in September 2010, which demonstrated flight control by use of fluidic devices. This enabled an entry into the Guinness Book of World records.

11.6 CONCLUSIONS

It is hoped that this chapter has given a brief sample of many of the exciting aircraft design projects that have spilled out of the creative minds of aircraft designers. There are many rewards, but also many pitfalls. Some of the problems have been noted in terms of the wrong requirements, projects overtaken by world events, over-ambitious projects or those which came at the wrong time. Other problems not already shown include political interference, too long a gestation period, which led to aircraft being too late to the marketplace or just plain bad luck.

Some of the more obvious problems can be reduced by paying much more attention to getting the specification right, and choosing an appropriate level of technology. Risk assessment

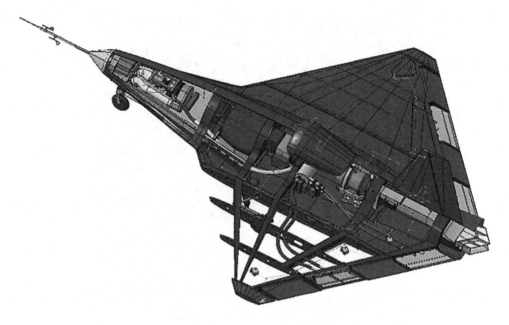

Fig. 11.23 CAD model of the Demon UAV.

Fig. 11.24 Cranfield Demon in flight.

can more easily be performed now by using computational methods to go further down the design cycle than was possible in the past. By this means, potential 'show-stoppers' can be isolated. There is a strong case for developing low-cost flying demonstrator programmes to investigate risky technologies – possible using remotely piloted vehicles.

The author unfortunately has no obvious answers to the whims of politicians, but wishes 'good luck' to designers of future aircraft.

APPENDIX A: USEFUL AIRCRAFT DESIGN DATA

A1 INTRODUCTION

The aircraft industry is littered with a plethora of aerospace terms and units, which leads to considerable confusion. The early part of this appendix aims to translate terms used by the English-speaking nations on either side of the Atlantic ocean.

Conversion tables are also provided to allow comparison between commonly used English or US units and their equivalent International System (SI) units.

Relevant, accurate, empirical aircraft design data are food and drink to an aircraft designer. It is important to learn from the past and use information about it as a guide for the future. Data comes from many sources, some more easy to acquire than others. Chapter 9 lists many data sources, but these still leave significant gaps in information. The author has accumulated much data over the years, and has extracted others from regular sources to present what is hoped will be useful information in a simple form for use in the early design processes. These data, by their nature, become obsolete after a few years, but it will be possible to update them as more information becomes available.

This appendix gives information in tabular or pictorial form in such areas as aircraft geometric, mass and performance data. It also gives information on powerplants, aerodynamics, structures, landing gear, interiors and weapons.

A2 US/UK NOMENCLATURE

Figure A2.1 shows the main components of a commercial aircraft, together with relevant names.
Other components, not shown:

US	UK
Empennage	Tail unit
Canard	Foreplane
Aluminum	Aluminium
Windshield	Windscreen
Zee-section (of stringer)	Zed-section

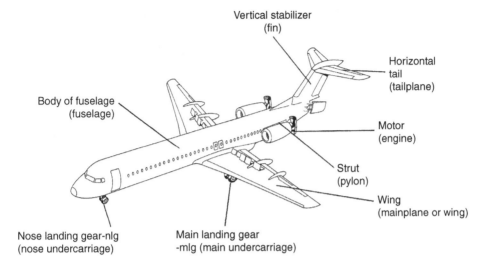

Fig. A2.1 Major component nomenclature – US and (British).

A3 UK AND US/SI CONVERSION TABLES AND AIRSPEED CHARTS

Many US companies use what they term 'English' units for dimensions, speeds, volumes etc. Most UK and other European companies and universities use SI units. The conversion chart in Table A3.1 should be useful in making conversions of aeronautical measures.

Speed, altitude and Mach number conversions often lead to difficulties, so Fig. A3.1 was produced to show the relationships between these values. Table A3.2 shows the values of the main atmospheric properties at sea level.

If

$$V_e = \text{Equivalent airspeed}\,(EAS),\ \text{True airspeed}\,(TAS) = \frac{EAS}{\sqrt{\delta}}$$

$$\delta = \text{density ratio} = \frac{\rho \text{ at a given altitude}}{\rho_0 \text{ at sea level ISA conditions}}$$

Mach no, $M = \frac{V_a}{a} = \frac{V_a}{\sqrt{\gamma g R T}}$

where V_a = true airspeed; γ = specific heat ratio; R = gas constant; T = ambient temperature.

Table A3.1   Conversion factors for widely used aeronautical units

Category	English (or US) units		SI units
Linear dimensions			
Inch	1 in	=	25.4 mm
Foot	1 ft	=	0.3048 m
Yard	1 yard	=	0.9144 m
Mile	1 mile	=	1.609 km
Nautical mile	1 nm	=	1.853 km
Areas			
Square inch	1 in^2	=	16.39 cm^2
Square foot	1 ft^2	=	0.0929 m^2
Volumes			
Cubic inch	1 in^3	=	16.39 cm^3
Cubic foot	1 ft^3	=	0.02832 m^3
UK gallon	1 gall(UK)	=	4.546 l
US gallon	1 gall(US)	=	3.785 l
Masses			
Pound	1 lb	=	0.4536 kg
Ton	1 ton	=	1016 kg
Slug	32.2 lb	=	14.6 kg
Densities			
Slug per cubic foot	1 slug/ft^3	=	515.4 kg/m^3
General density	1 lb/in^3	=	27.68 kg/l
	1 lb/ft^3	=	16.02 kg/m^3
Forces, pressures, stresses			
Pound force	1 lbf	=	4.448 N
Pound per square inch	1 lbf/in^2	=	6.895 kN/m^2
		or	68.95 mb
Ton per square inch	1 ton f/in^2	=	15.44 MN/m^2
Pound mass per square foot	1 lb/ft^2	=	4.8825 kg/m^2
Pound force per square foot	1 lbf/ft^2	=	47.88 N/m^2
1 atmosphere	14.7 lbf/in^2	=	0.3045 m/s
Velocities			
Feet per second	1 ft/sec	=	0.3048 m/s
Mile per hour	1 mph	=	0.447 m/s
		=	1.609 km/h
Knot	1 kt	=	0.5148 m/s
		=	1.853 km/h

see Fig. A3.1 for conversions between airspeeds and Mach numbers

Aerospace terms			
Specific fuel consumption for jet engines = pound of fuel per pound of thrust per hour (s.f.c)	1 lb/lbf/h	=	28.32 mg/N/s

Table A3.1   (*cont.*)

Category	English (or US) units		SI units
Specific fuel consumption – turbo-prop	1 lb/h/hp	=	168.9 µg/J
Specific fuel consumption – piston	1 lb/h/hp	=	168.9 µg/J
Horsepower	1 hp	=	745.7 W

SI symbols: m = metre, mm = millimetre, cm = centimetre, km = kilometre, l = litre, g = gram, kg = kilogram, mg = milligram, µg = microgram, MN = meganewton, N = newton, b = bar, mb = millibar, s = second, J = joule, W = watt

Table A3.2   Atmospheric properties at sea level

Standard sea level conditions		English (or US) units	SI units
Pressure	(P_0)	2116.221 bf/ft^2	101 325.0 N/m^2
		(14.6961 bf/in^2)	(1013.25 millibar)
Temperature	(t_0)	49°F	15°C
	(T_0)	518.67 R	288.16 K
Density	(ρ_0)	0.002 376 9 slugs/ft^3	1.2250 kg/m^3
Velocity of sound	(a_0)	1116.45 ft/s 661.48 knots	340.294 m/s
Kinematic viscocity	(ν_0)	1.5723 × 10^{-4} ft^2/s	1.4607 × 10^{-5} m^2/s
Acceleration due to gravity	(g_0)	32.1741 ft/s^2	9.806 65 m/s^2

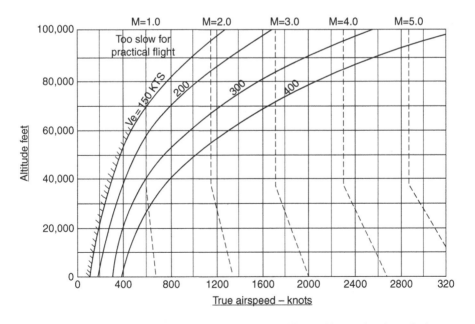

Fig. A3.1 Speed, altitude and Mach number relationships and international standard atmosphere (ISA) sea level conditions.

A4 AIRCRAFT LEADING DATA TABLES

This section (Tables A4.l–A4.ll) is a compilation of a range of data that should be useful for the early conceptual design process and for use as a 'reality check' as the design progresses. It is crucial to know the performance and characteristics of current aircraft, when designing competitors. The format of the data was chosen to be directly applicable for use in parametric analyses. Sources include:

(i) Several editions of Jane's *All The World's Aircraft* [6].

(ii) Extracts from compilations in *Flight International* magazine.

(iii) A Cranfield research thesis by R. Hewson produced in 1993 [45].

(iv) Inputs from manufacturers' catalogues.

Some of the data for combat aircraft have been reproduced from Hewson's thesis as Fig. A4.1 and give some useful information for the early stages of wing design.

Although care has been exercised in compiling this information, aircraft and engine performance often changes with time and therefore aircraft or engine manufacturers' data should be consulted for precise, current, numbers. The data in these tables contain information current at the end of 1996.

Table A4.1 Narrow-body jet transports – leading dimensions

Information on more recent types can be found on the relevant company websites. Important newcomers or projects include the Airbus A320NEO, Boeing 737MAX, Bombardier C-Series, EMBRAER E-jets, COMAC ARJ-21 and 919, Mitsubishi MRJ and the Sukhoi Superjet 100.

Manufacturer	Model	Max. passengers high density	Dimensions							Engines		
			Span (b) (m)	Fuselage width (m)	Fuselage length total (m)	Height (m)	Wing gross area (S) (m²)	Wing aspect ratio	Sweep quarter chord (°)	No. engines	Engines sea level thrust (D(kN))	Engine model
Airbus Industrie	A319	153	34.10	3.95	33.84	11.76	122	9.5	25	2	97.5 104	CFM565A4 IAEV2522-A5 CFM565A5 IAEV2524-A5
	A320–200 A320NEO A320NEO	180	34.10	3.95	37.57	11.76	122	9.5	25	2	111	CFM56-5A1 CFM56 CFM LEAP-X PW1127G
Avro International	RJ70	94	26.21	3.56	23.8	8.61	· 77.3	8.89	15	4	27.28	LF507
Boeing	B737–200	130	28.30	3.76	30.53	11.28	91	8.8	25	2	72	P&WJT8D–15Al
	B737–500	132	28.90	3.76	31.00	11.10	91	8.8	25	2	89	CFM56–3139
	B737–300	149	28.90	3.76	33.40	11.10	91	8.8	25	2	99	CFM56–3B–2/–3C–1
	B737–Max9	180		3.76						2		CFM LEAP1-B
	B757–200	189	38.05	3.76	47.32	13.56	181	8.0	25	2	178	RB211–535E4
Fokker	70	79	28.08	3.3	30.91	8.51	93.50	8.4	17.4	2	61.6	R-RTAYMK620–15
Illyushin	11–62MK	174	43.02	4.1	53.1	12.35	280	6.6	32.3	4	107.9	Soloviev D–30KU
McDonnell	MD–81	139	32	3.4	39.75	9.3	112.3	9.6	24.5	2	92.71	P&WJT8D–217B/C
Douglas	MD–83	172	32.9	3.4	45.0	9.0	112.3	9.6	24.5	2	96.5	P&WJT8D–219
	MD82/88	172	32.9	3.4	45.0	9.0	112.3	9.6	24.5	2	92.7	JT8D–217AC
Tupolev	TU204	240	42.0	3.8	48.0	13.88	184.1	9.6	28	2	157	Soloviev PS–90A
Yakovlev	YAK–46	162	36.25	3.8	40.38	9.8	120	11	25.0	2	114	Lotariev D–27
	YAK–42M	168	36.25	3.8	40.38	9.8	120	11	25	3	73.6	Lotariev D–436M

Table A4.2 Narrow-body jet transports – weights and performance

Model	Weights					Payload range		Maximum cruise				Cruise – long range		
	Taxi-ing	Max take-off (W_G)	Landing (W_L)	Empty operating	Fuel (max)	Max payload	Range at max payload	Speed	Speed	Altitude	Fuel consumption	Speed	Altitude	Fuel consumption
	(kg)	(kg)	(kg)	(kg)	(litres)	(kg)	(kM)	(knot)	(Mach)	(ft)	(kg/h)	(knot)	(ft)	(kg/h)
A319	64 400	64 400	61 000	39 500	23 860	17 400	2525	487	0.82	28 000	3150	450	37 000	1950
A320–200	73 900	73 500	64 500	41000	23 860	20 000	4240	487	0.82	28 000	3200	448	37 000	2100
RJ70	38 328	38102	37 875	23 781	11 728	8 650	2046	432	0.73	29 000	2150	356	29 000	1635
737–200	52 620	52 390	47 600	27 520	19 531	15710	2868	488	0.88	25 000	4259	420	35 000	2327
737–500	52 620	52 390	49 900	30 960	20 105	15 530	2519	492	0.82	26 000	3574	429	35 000	2100
737–300	56 700	56 470	51 710	31869	20 105	16 030	2923	491	0.82	26 000	3890	429	35 000	2250
737–400	63 050	62 820	54 880	34 470	20 105	17 740	3611	492	0.82	26 000	3307	430	35 000	2377
757–200	100 245	99 792	89 800	58 248	42 320	26 090	5890	505	0.88	31 000	5350	459	39 000	2885
F70	38 325	38100	35 830	23 100	13 365	9 555	4300	461	0.77	26 000	2391	401	35 000	1475
IL – 62MK	NA	165 000	105 000	71600	105 300	23 000	7800	496	0.77	26 200	NA	460	36 600	NA
MD81	63 950	63 500	58 000	33 253	22 106	17619	3450	499	0.84	27 000	3924	439	35 000	2458
MD83	73 030	72 580	63 280	36 620	26 495	18721	4387	499	0.84	27 000	4027	439	35 000	2518
MD82/88	68 270	67 812	58 970	35 630	22 106	18 802	3445	499	0.84	27 000	4077	439	35 000	2560
TU204	93 850	93 500	86 000	56 500	30 000	21000	2500	458	0.9	40 000	3270	NA	NA	NA
YAK46	61 600	61300	56 750	31300	18 590	17 500	1800	432	0.8	36 000	1570	415	36 000	1500
YAK42M	63 300	63 300	56 000	38 500	18 200	16 500	1500	432	0.78	36 400	2300	416	36 400	2030

Table A4.3 Narrow-body jet transports – field performance and parametric ratios

	FAR field length (maximum weight)		Aircraft maximum lift coefficient			Landing gear		W_G/S		Parametric ratios		
	Take-off field ISA, SL (m)	Land field ISA, SL (m)	Take-off	Approach	Landing	Track (m)	Wheel-base (m)	kg/ m^2	lb/ ft^2	T_o/W_G	W_L/W_G	Fuselage length diameter
A319	1750	1350		2.6	2.8	7.6	11.4	528	108	0.329	0.947	8.567
A320–200	2180	1440		2.6	2.8	7.5	12.6	602	123	0347	0.877	9.511
RJ70	1440	1170		3.3	3.5	4.72	10.09	493	100	0.292	0.994	6.68
737–200	1829	1350				5.25	11.4	576	118	0.28	0.909	8.12
737–500	1518	1362				5.25	14.1	576	118	0.346	0.952	8.24
737–300	1600	1396		2.4	3.3	5.25	12.4	621	127	0.357	0.916	8.883
737–400	1935	1506				5.25	14.3	690	141	0.341	0.874	9.681
757–200	1880	1415		1.9	2.5	7.32	18.29	551	113	0.364	0.90	12.59
F70	1391	1208				5.04	11.54	407	83	0.33	0.94	9.367
IL–62MK	3300	2500				6.8	24.5	589	121	0.267	0.636	12.95
MD81	1850	1451		2.2	3.0	5.1	19.2	565	116	0.298	0.913	11.69
MD83	2551	1585				5.1	22.0	646	132	0.271	0.872	13.24
MD82/88	2274	1500				5.1	22.0	604	124	0.279	0.87	13.24
TU204	2765	2900				10.7	23.3	508	104	0.342	0.92	12.63
YAK46	1500	1670				5.63	14.78	511	105	0.379	0.926	10.626
YAK42M	1950	2020				5.63	14.78	528	108	0.356	0.885	10.626

Table A4.4   Wide-body jet transports – leading dimensions

Information on more recent types can be found on the relevant company websites. Important newcomers or projects include the Airbus A350XWB, A380, Boeing 747-9, 777X and 787

Manufacturer	Model	Max passengers high density	Dimensions							Engines		
			Span (b) (m)	Fuselage width (m)	Fuselage length total (m)	Height (m)	Wing gross area (5) (m²)	Wing aspect ratio	Sweep quarter chord (0)	No. engines	Engines sea level thrust (T) (kN)	Engine model
Airbus Industrie	A310–200	280	43.9	5.64	45.13	15.81	219	8.8	28.0	2	237	GE CF6-8-C2A2
	A300–600	375	44.84	5.64	53.3	16.53	260	7.7	28.0	2	262	GE CF6-80C2A1
	A330–300	440	60.3	5.64	63.65	16.90	362	10.0	30.0	2	300	R-R Trent
	A340–200	440	60.3	5.64	59.4	16.8	362	10.0	30.0	4	139	CFM56-5C2
Boeing	767–200	290	47.57	5.03	48.51	15.85	283	8.0	31.5	2	22	PW4050
	777–200	440	60.9	6.2	62.78	18.4	428	8.68	31.6	2	331	Various
	747–200C	550	59.6	6.6	68.63	19.3	512	6.94	37.5	4	244	P&W JT9D–7R4G2
	747–400	660	64.3	6.6	68.63	19.3	525	7.87	37.5	4	258	R-R RB211–524G
Illyushin	IL–96–300	300	57.66	6.08	51.15	17.6	392	9.5	30.0	4	156	Soloview PS–90A
Lockheed	L1011–250 TRISTAR	400	47.35	5.97	54.17	18.9	322	6.95	35.0	3	222	R-R RB211–524B
McDonnell Douglas	DC 10–30	380	50.42	6.02	51.97	17.7	338	7.5	35.0	3	234	GE CF6–50C2
	MD–11	405	51.77	6.02	58.65	17.6	339	7.5	35.0	3	274	GE CF6–80C2

Table A4.5 Wide-body jet transports – weights and performance

Model	Weights					Payload range		Maximum cruise				Cruise – long range		
	Taxi-ing (kg)	Max take-off (W_G) (kg)	Max landing (W_L) (kg)	Empty operating (kg)	Fuel (max) (litres)	Max payload (kg)	Range at max payload* (kM)	Speed (knots)	Speed (Mach)	Altitude (ft)	Fuel consumption (kg/h)	Speed (knots)	Altitude (ft)	Fuel consumption (kg/h)
A310–200	142 900	142 000	123 000	79 450	54 920	33 550	5350	484	0.84	35 000	4 740	459	37 000	3 770
A300–600	165 900	165 000	138 000	87 600	62 000	42400	5 250	480	0.84	31 000	5 160	454	35 000	4 300
A330–300	212 900	212 000	174 000	117 700	97170	45 470	8 785*	500	0.86	33 000	5 000	465	39 000	4 700
A340–200	257 900	257 000	181 000	119 600	138 600	49 400	11 800	500	0.86	33 000	7 180	475	39 000	5 400
767–200	152 860	151 950	126 099	80 780	63 220	33 970	5 800	492	0.83	39 000	5 370	457	39 000	3 674
777–200	230450	229 500	201 850	136100	117 340	54 450	5 460	499	0.83	39 000	–	476	39 000	–
747–200C	379210	377 850	285 830	180 850	198 350	92 310	8 220	507	0.88	35 000	12 900	484	35 000	10 700
747–400	395 990	394 630	285 760	178810	204 350	63 870	12 870	507	0.88	35 000	11 230	490	35 000	9 840
IL96–300	N/A	216 000	175 000	117 000	150 000	39 990	7170	470	0.83	30 000	–	460	39 000	–
L1011–250	232 240	231 330	166 920	113 290	119 780	49 030	8 980	518	0.89	33 000	7 620	483	33 000	6 940
DC10–30	265 000	265 000	182 800	120 900	146 290	45 910	9 950	530	0.88	25 000	9 743	490	30 000	7 310
MD–11	285 080	283 720	195 000	131000	146 290	55 570	11 100	511	0.87	27 000	8 970	473	35 000	7 060

* Range with 335 passengers and baggage.

Table A4.6 Wide-body jet transports – field performance and parametric ratios

Take-off field ISA, SL (m)	FAR field length (max weight)		Aircraft max lift coefficient		Landing gear		W_G/S		Parametric ratios		Fuselage length diameter
	Land field ISA, SL (m)	Take-off	Approach	Landing	Track (m)	Wheel-base (m)	(kg/m²)	(lb/ft²)	T_o/W_G	W_L/W_G	
A310–200	1860	1480	2.4	3.1	9.6	15.2	648	133	0.340	0.866	8.0
A300–600	2240	1532	2.4	3.0	9.6	18.6	635	130	0.324	0.836	9.45
A330–300	2620	2150	2.1	2.6	10.49	–	586	120	0.288	0.821	11.29
A340–200	2765	2900	2.5	2.6	10.49	23.3	710	145	0.220	0.704	10.53
767–200	1880	1415	2.2	2.4	7.32	18.29	537	110	0.299	0.830	9.64
777–200	2135	1610	–	–	10.97	25.88	538	110	0.294	0.88	10.13
747–200C	3190	2100	–	–	11.0	25.6	741	152	0.263	0.724	10.4
747–400	3320	2130	1.9	2.4	11.0	25.6	754	154	0.267	0.724	10.4
IL96–300	2760	1980	–	–	10.4	20.07	551	113	0.294	0.81	8.41
L1011–250	2990	2040	–	–	10.97	21.34	718	147	0.293	0.722	9.07
DC10–30	2996	1820	–	–	10.67	22.07	784	161	0.27	0.69	8.63
MD–11	2930	1970	2.2	2.7	10.57	24.61	837	171	0.295	0.687	9.74

Table A4.7 Regional turbo-props – leading dimensions

Manufacturer	Model	Max passengers high density	Span (b) (m)	Fuselage Width (m)	Fuselage length total (m)	Height (m)	Wing gross area (S) (m²)	Wing aspect ratio	Propellor diameter (m)	No. engines	Engines sea level power (kW)	Engine model
Aritonov	AN-32	50	29.2	2.9	23.8	8.75	75	11.7	4.7	2	3812	IvchenkoA-20D
Avions de Transport Regional	ATR-42-300	50	24.57	2.87	22.67	7.59	54.5	11.08	3.96	2	1342	PWC PW120
	ATR-72	74	27.05	2.87	27.17	7.65	61.0	12.0		2	1611	PWC PW124B
BAE SYSTEMS	Jetstream Super 31	19	15.85	1.98	13.4	5.38	25.2	9.95	2.69	2	760	GarrettTPE331-12UAR
	Jetstream 41	29	18.29	1.98	18.25	5.74	32.59	10.26	2.9	2	1118	GarrettTPE 331-14 GR/HR
	ATP	72	30.63	2.67	25.46	7.54	78.32	11.984	4.19	2	1978	P&WPW126A
CASA/IPTN	CN235-100	44	25.81	2.9	20.9	8.177	59.1	11.3	3.35	2	1305	GE CT7-C
de Havilland Canada	DASH8-300A	56	27.43	2.69	24.28	7.49	56.21	13.39	3.96	2	1864	PW123B
Embraer	Brasilia EMB-120	30	19.78	2.28	18.73	6.35	39.43	9.9	3.2	2	1342	PW118
Fairchild	Metro 23	20	17.37	1.68	17.4	5.08	28.71	10.5	2.69	2	820	TPE331-12UA
Fokker	50-100	58	29	2.7	25.25	8.317	70.0	12.0	3.66	2	1864	PWC PW125B
Saab	340B	37	21.44	2.31	19.73	6.87	41.81	11.0	3.35	2	1305	GE CT7-9B
	2000	58	24.76	2.31	27.03	7.73	55.7	11.0	3.81	2	3096	Allison GMA 2100
Shorts	330-200	30	22.76	2.24	17.69	4.95	42.1	12.3	2.82	2	893	PWC PT6A-45R

Table A4.8 Regional turbo-props – weights and performance

Model	Weights				Payload range		Max cruise		Cruise – long range	
	Max take-off (W_o) (kg)	Max landing (W_L)(kg)	Empty operating (kg)	Fuel (max) (kg)	Max payload (kg)	Range at max payload (km)	Speed (knot)	Altitude (ft)	Speed (knot)	Altitude (ft)
AN–32	27 000	25 000	17 308	5445	6700	1200	286	26 250	254	26 250
ATR–42	16 700	16 400	10 285	4500	4915	1946	250	17 000	243	25 000
ATR–72	21500	21 350	12 500	5000	7200	1195	284	25 000	248	25 000
J31	7 350	7 080	4 578	1372	1805	1192	264	15 000	244	25 000
J41	10 433	10 115	6 350	2703	3039	1263	295	20 000	260	20 000
ATP	22 930	22 250	14 202	5080	7026	630	266	13 000	236	18 000
CN235–100	15 100	14 900	9 800	4230	4000	834	248	15 000	–	–
Dash8–300A	18 642	18 144	11 657	2576	5216	1538	287	15 000	–	–
Brasilia	11500	11 250	7 465	2600	3039	1019	300	25 000	260	25 000
Metro	7 484	7110	4 085	1969	1710	2174	288	11 000	–	–
F50	19 990	19 500	12 520	4123	6080	2055	282	–	–	–
S340	12 927	12 700	8 035	2581	3758	1520	282	15 000	252	25 000
S2000	22 000	21 500	13 500	4165	5900	2279	366	25 000	300	31 000
S330	10 387	10 251	6 680	2032	2653	876	190	10 000	160	10 000

Table A4.9   Regional turbo-props – field performance and parametric ratios

| Model | FAR field length (max weight) | | Landing gear | | W_G/S | | Parametric ratios | | |
	Take-off field ISA, SL (m)	Landing field ISA, SL (m)	Track (centreline of legs) (m)	Wheel-base (m)	(kg/m²)	(lb/ft²)	P/W_G	W_L/W_G	Fuselage length diameter
AN–32	1200*	450**	7.9	7.65	360	74	0.282	0.926	8.21
ATR–42	1040	1030	4.1	8.78	306	63	0.161	0.982	7.9
ATR–72	1408	1210	4.1	10.70	352	72	0.15	0.993	9.47
J31	1569	1364	5.94	4.6	292	60	0.207	0.963	6.77
J41	1523*	1250**	6.1	7.32	320	65	0.214	0.97	9.22
ATP	1463	1128	8.46	9.7	293	60	0.173	0.97	9.54
CN235–100	1400	1165**	3.9	6.92	255	52	0.173	0.987	7.21
Dash 8	1067	1006	7.87	10.01	332	68	0.20	0.973	9.02
Brasilia	1420	1370	6.58	6.97	292	60	0.233	0.978	8.21
Metro	1414*	848**	4.57	5.83	261	53	0.219	0.95	10.36
F50	940	1020	7.2	9.7	286	59	0.186	0.975	9.35
S340	1271	1049	6.71	7.14	309	63	0.202	0.982	8.54
S2000	1360	1250	8.23	10.97	395	81	0.281	0.977	11.7
S330	1042	1030	4.24	6.15	247	51	0.172	0.987	7.9

* Take-off to 15 m.
** Landing run.

Table A4.10   Combat aircraft and trainers – leading dimensions

Manufacturer	Aircraft type	Category	Length (m)	Wing sweepback leading edge (°)	Wing sweepback quarter-chord 1/4 (°)	Wing thickness (inner/outer) t/c (%)	Wing span, b (m)	Wing area, S (m²)	Wing aspect ratio	Wing taper ratio λ
BAE SYSTEMS	Hawk 200	F	11.33	26.0	21.5	(10.9/9.0)	9.39	16.7	5.30	0.34
Sepecat	Jaguar	A	15.52	48.0	40.0	(6.0/5.0)	8.69	24.2	3.12	0.32
Fairchild	A-10	A	16.26	5.0	0.0	(16.0/13.0)	17.53	47	6.54	0.65
Lockheed	F-117A	A	20.08	67.5	64.0	N/A	13.2	105.9	1.65	N/A
BAE SYSTEMS/ McD-D	Harrier II	V/A	14.12	36.0	24.0	(11.5/7.5)	9.24	21.4	4.00	0.28
	YAK-38	V/F	15.5	46.0	40.0	N/A	7.31	18.5	2.89	0.26
	YAK-141	V/F	18.3	39.0	35.0	N/A	10.09	31.3	3.26	0.28
McD-Douglas	F-15E	F	19.43	45.0	38.7	(6.6/3.0)	13.05	56.5	3.01	0.23
Lockheed	F-16C	F	15.03	40.0	32.0	4.0	9.45	27.9	3.00	0.21
McD-Douglas	F/A-18E	F	18.31	26.0	20.0	(5.0/3.5)	11.43	46.5	3.52	0.37
Northrop	F-20	F	14.42	33.0	25.0	4.8	8.14	18.6	3.56	0.19
Lockheed	YF-22	R	19.56	48.0	37.1	N/A	13.11	78.0	2.20	0.14
Northrop	YF-23	R	–	40.0	22.8	N/A	13.29	87.8	2.01	0.09
Sukhoi	Su-27	A	21.94	42.0	37.0	N/A	14.69	60.6	3.57	0.28
	MiG-29	F	17.32	42.0	35.0	N/A	11.37	38.0	3.50	0.23
Rockwell	X-31A	R	13.21	N/A	48.1	5.5	7.25	21.0	2.51	N/A
	IAI Lavi	F/A	14.39	54.0	47.0	N/A	8.78	33.1	2.33	N/A
	Eurofighter	F	14.5	53.0	44.0	N/A	10.52	50.0	2.21	0.07
SAAB	JAS39	F	14.1	52.0	43.0	N/A	8.02	27.9	2.29	0.18
Dassault	Rafale D	F	15.3	47.0	39.0	N/A	10.45	46.0	2.37	0.20
Panama	Tornado	A	18.68	25.0–67.0	N/A	N/A	8.6–13.9	26.7	2.78–7.27	N/A
Grumman	F-14A	F	19.1	20.0–68.0	15.5–63.5	(10.0/5.0)	(11.6–19.54)	52.5	2.07–7.28	0.25–0.31
BAE SYSTEMS	Hawk 100	T	11.68	26.0	21.5	10.9/9.0	9.37	16.7	5.3	0.34
Sai Marchetti	S211	T	9.31	–	15.3	15.0/13.0	8.43	12.6	5.1	0.465
Pilatus	PC-9	T	10.175	–	1	(OUTBRD) 15.0/12.0	15.0/12.0	10.124	16.29	6.3

Categories: A = attack, F = fighter, V = vertical take-off on landing, R = research, T = trainer.

Table A4.11 Combat aircraft and trainers-engines, weights and performance

Column groups: *Uninstalled engine performance* (No. Engines – (kN) max afterburner thrust); *Weights* (Empty – Max payload); *Performance* (Typical combat radius – Typical landing run).

Aircraft type	No. Engines	Model	All engines max static thrust (lbf)	(kN)	All engines max afterburner thrust (lbf)	(kN)	Empty (lb)	(kg)	Max internal fuel (lb)	(kg)	Max take-off (lb)	(kg)	Max payload (kg)	Typical combat radius (km)	Max Mach no. at altitude	Ceiling (km)	Typical take-off run (m)	Typical landing run (m)
Hawk 200	1	RR Adour 871	5 845	26	N/A	–	9 943	4450	3 000	1360	20 065	9100	4 128	1072	0.9	15.25	630	598
Jaguar	2	RR Adour	10 640	47.3	16 080	71.5	15 432	7 006	N/A	–	34 612	15 710	4 500	815	1.6	–	580	470
A-10	2	GETF34-100	18 130	80.6	N/A	–	24 959	13 330	10 700	4 860	50 000	22 100	7 250+	480	0.6	–	1173	652
F-117A	2	GE F404	21 600	96.1	N/A	–	30 000	13 620	N/A	–	52 500	23 840	2 270	1112	1.1	–	–	–
Harrier II	1	RR Pegasus	21 450	95.4	N/A	–	14 860	6 746	7 759	3 520	31 000	14 075	5 900	1110	1.0	13.7	0–500	0
YAK-38	1+2[a]	Tumansky R-27V-300	15 300	68.1	N/A	–	16 500	7 490	N/A	–	25 795	11 710	–	185	1.0	12.0	0	0
YAK-141	1+2[b]	Soyuz R-79	19 840	88.2	34 170	152	N/A	–	N/A	–	42 990	19520	2 600	1400	1.7	15.0+	0	0
F-15E	2	P&WF100	N/A	–	58 000	258	31 700	14 390	13 123	5 960	81 000	36 775	14 380	1850+	2.5	18.3	274	840
F-16C	1	P&WF100-229	N/A	–	29 000	129	19 020	8 635	6 846	3110	42 300	19 200	7618	925+	2.0	18.0	760	760
F/A-18E	2	GE F404	N/A	–	44 000	195	30 600	13 890	14 400	6 540	66 000	29 960	–	–	1.8	–	–	–
F-20	1	GEF404	N/A	–	18 000	80	13 150	5970	5 050	2 290	28 000	12 710	4080+	556	2.0	17.31	1080	655
YF-22	2	P&WF119	N/A	–	70 000	311	31 000	14 070	22 000	9 990	60 000	27 240	–	–	2.2	15.24	–	–
YF-23	2	P&WF119	N/A	–	70 000	311	37 000	16 800	24 000	10 900	64 000	29 060	–	–	2.0	–	–	–
Su-27	2	LyulkaAL-31F	N/A	–	55 114	245	N/A	–	N/A	–	49 600	22 520	17 700	2000	2.4	19.0+	–	–
MiG-29	2	Tumansky R-33D	22 220	98.8	36 600	163	24 030	10 910	N/A	–	40 785	18 520	8 175	1150	2.3	15.0+	–	–
X-31A	1	F404	N/A	–	16 000	71.2	11 410	5 180	4 136	1880	15 935	7235	–	–	2.3	12.0+	457	823
IAI Lavi	1	P&WPW1120	–	–	20 620	91.7	16 024	7275	6 000	2 725	42 500	19 295	7 257	1850	1.9	–	–	–
Eurofighter	2	Eurojet EJ-200	26 980	120	40 500	180	21 495	9760	8818	4 005	46 297	21 020	9 750	–	2.0	–	300	700
JAS39	1	VFARM12	12 140	54	18 100	80.5	14 600	6630	5 000	2 270	27 500	12 485	8 000	–	2.0	18.0	400	5000
Rafale D	2	Snecma M88-2	N/A	–	39 116	174	19 973	9070	N/A	–	47 399	21 520	8 000	1090+	2.0	–	400	–
Tornado	2	Turbo union RB 199	17 300	77	32 150	143	30 620	13 901	10 251	4 655	61620	27 975	14 500	740+	2.2	21/3 +	670	370
F-14A	2	P&W TF-30	28 000	124.5	46 200	206	40 104	18 210	16 200	7 355	74 349	33 755	6 600	–	2.3	15.2+	427[c]	884[c]
HAWK 100	1	RR Adour 871	5 845	26	–	–	9 692	4400	2 872	1304	20 045	9100	3 000	781	0.88	13.55	640	605
S211	1	PWCJT-15D-4C	2 500	11.1	–	–	4 075	1850	1370	622	6 057	2 750	660	556	0.6	12.2	390	361
PC-9	1	PWC PT6A-62	708 KW	–	–	–	3711	1685	–	–	7 050	3 200	–	887 (Range)	–	11.58	227	530

[a] Two vertical take-off Rybansk lift engines of 6725 lbf/(30 kN) each.

[b] Two vertical take-off Rybansk engines of 9390 lbf (41.8 kN) each.

[c] Minimum distance.

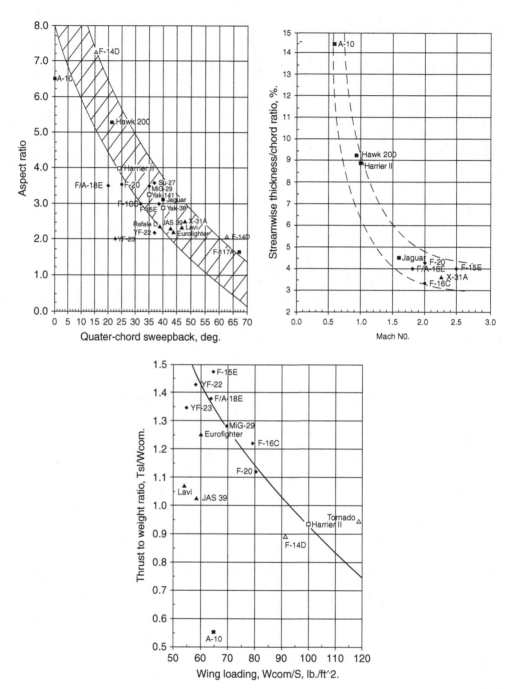

Fig. A4.1 Current combat aircraft major wing parameters.

A5 POWERPLANT DATA

This section is a summary of a wide range of data that should be useful for aircraft conceptual design purposes. The data are those that affect the aircraft, but there is no information about the details of the engine cycle design, as that is not the function of this section. The information is mainly in the form of summary tables for: civil turbo-fans; civil/military turbo-props; military turbo-fans/jets.

There is also information giving more detail about representative civil and military engines, and limited information about piston engines.

The primary data sources were:

(i) Several editions of Jane's *All The World s Aircraft* [6].
(ii) Extracts from compilations in *Flight International*.
(iii) Information from a European engine company, published in Cranfield theses with permission.
(iv) Several engine manufacturers' catalogues.
(v) Results from running Cranfield University's engine design program, Turbo-Match.

A5.1 Engine Data Tables

Information from the above sources was compressed and presented in a consistent format in the following tables for civil turbo-fan engines (Table A5.1); turbo-prop engines (Table A5.2); and military turbo-jet and turbo-fan engines (Table A5.3).

A5.2 Detailed Data of Project Engines

A5.2.1 The T-91 Project Turbo-Fan Engine
This engine was designed by an engine company's project office to support the Cranfield T-91 advanced trainer project [46].

A5.2.1.1 Engine Description
The engine proposed is a two-spool, medium bypass (typically 1.5:1) engine, rated for trainer operating cycles. Maximum design stator outlet temperature is only reached during hot-day take-off runs and at the maximum sea level ISA-day speed of M 0.9. There is no time restriction on any other performance levels.

The engine carcass dimensions are shown in Fig. A5.1. These have been derived by scaling the original drawing provided by the manufacturer. Also, a 120 mm jet pipe extension has been inserted between the turbine exit and the exhaust nozzle, as that was required by acoustic fatigue considerations at the rear fuselage tail boom section. This jet pipe has increased the total engine mass given by the manufacturer from 388 kg to about 400 kg (an increase of 12 kg). On the given

Table A5.1 Powerplant data: civil turbo-fan engines

Manufacturer	Engine type	Sea level static thrust (ISA)		SLST specific fuel consumption (lb/lbf/h)	Bypass ratio	Dry mass consumption		Maximum typical cruise thrust (lbf × 100)/(Mach)/(ALT 1000 ft)	Typical cruise specific fuel consumption (lb/lbf/h)	Length		Fan diameter	
		(kN)	(lbf × 1000)			(kg)	(lb)			(m)	(in)	(m)	(in)
Allied Sigma Engines	TFE 731-5	19.1	4.3		3.34	409	900	0.98/0.8/40	0.8	1.66	65.4	0.75	29.7
BMW Rolls-Royce	BR710	65	14.75		4.0	209	460	3.58/0.8/35	0.64	5.1^a	200.8	1.11	43.7
BMW Rolls-Royce	BR715	88.8	19.9		4.7			4.4/0.8/35	0.62	5.2^a	205	1.34	52.8
CFE	738-1	25.5	5.72	0.39	5.3			1.3/0.8/40	0.645	2.52	99.2	0.9	35.5
CFM International	CFM 56-3B	89	20	0.38	5.0	1943	4280	4.86/0.8/35	0.67	2.36	93	1.52	59.8
CFM International	CFM 56-5C	139	31.2	0.32	6.6	2490	5494	6.9/0.8/35	0.57	2.62	103	1.84	72.4
General Electric	CF34-3A1	41	9.22	0.36						2.615	103	1.24	48.8
General Electric	CF6-80-C2	233	52.5	0.32	5.05	4250	9360	11.3/0.85/35		4.09	161	2.36	92.9
General Electric	GE90-85B	377	84.7		8.4	7080	15 600			4.88	192	3/35	132
International Aero. Eng.	V2500-A1	111.2	25		5.4	2304	5074	5.07/0.8/35	0.58	3.2	126	1.6	63
Pratt & Whitney	JT8D-200	77	17.4	0.51	1.77	2070	4524	4.95/0.8/30	0.724	3.92	154	1.17	46
Pratt & Whitney	PW4052-4460	231	52	0.34	5.8	4268	9400	-/0.8/35	0.537	3.37	133	2.38	94
Pratt & Whitney	PW 4084	376	84		6.41	6606	14 550			3.37	133	2.84	112
Pratt & Whitney Canada	PW 305B	23.6	5.23	0.39	4.5	427	940	1.15/0.8/40	0.68	2.07	81.5	0.8	31.5
Rolls-Royce	TAY611	61.6	13.85		3.04	1340	2951	2.55/0.8/35	0.69	2.4	94.5	1.1	43.7
Rolls-Royce	RB211-535E4	178	40.1		4.3	3300	7264	8.5/0.8/35	0.60	2.99	117.7	1.88	74
Rolls-Royce	RB211-524G	258	58		4.3	4390	9670	11.8/0.85/35	0.57	3.18	125	2.19	86.2
Rolls-Royce	Trent 768	300	67.5		5.0	4785	10 550	11.5/0.82/35	0.565	3.9	153.5	2.47	97.2
Rolls-Royce	Trent 890	406	91.3		5.74	6000	13 100	13/0.83/35	0.557	4.3	169	2.79	110
Textron Lycoming	LF507-1F	31.1	7	0.406	5.6	629	1385			1.48	58.3	1.2	47.2
Williams Rolls-Royce	FJ44	8.2	1.9	0.48	3.28	202	445	0.6/0.7/30	0.75	1.18	46.5	0.6	24

^{a} Nacelle dimensions.

Table A5.2   Powerplant data: turbo-prop engines

Manufacturer	Engine type	Take-off shaft horsepower	SFC Take-off (lb/h/ESHP)	Dry mass (inc. gearbox)		Length		Maximum width	
				(kg)	(lb)	(m)	(in)	(m)	(in)
Allied Signal Engines	TPE 331-12	1100	0.52			1.1	43.3	0.53	21
Allied Signal Engines	TPE331-14GR	1650	0.51			1.35	53	0.58	23
Allison	AE2100A	3690	0.415			1.96	77	0.67	264
Allison	T56	4920	0.50			3.71	146	0.99	39
General Electric	CT7-9	1740	0.48	295	650	2.44	96	0.74	29
Pratt & Whitney Canada	PT6A-25C	550	0.63			1.57	62	0.48	19
Pratt & Whitney Canada	PW124B	2950	0.454			2.06	81	0.84	33
Pratt & Whitney Canada	PW127	3300	0.449			2.06	81	0.84	33
Rolls-Royce/Snecma	Tyne MK21	6100				2.76	109	0.84	43

Table A5.3   Powerplant data: military turbo-jet and turbo-fan engines

Manufacturer	Engine type	Sea level static thrust reheat (kN)	(lbf × 1000)	(SLST) Sea level static thrust – dry (kN)	(lbf × 1000)	SLST dry specific fuel consumption (lb/lbf/h)	Bypass ratio	Basic engine mass (kg)	(lb)	Length (m)	(in)	Diameter (m)	(in)
Allied Signal Engines	TFE 1042-70	42.1	9.5	27	6.06	0.8 (dry)	0.3	617	1360	2.88	113	0.6	24
Eurojet	EJ200	90	20				0.4	Approx. 1000	Approx. 2000	4.0	158	0.74	29
General Electric	F404-402	79	17.7	53	11.9			1035	2280	4.04	159	0.89	35
General Electric	F110-129	131.3	29.5	78.3	17.6		0.76			4.62	182	1.18	47
Klimov	RD-33	81	18.3	50	11.24	0.76 (dry)		1055	2324	4.17	164	1.0	39
Pratt & Whitney	F100-229	129.5	29.1	79.2	17.8	2.05 (wet)	0.36	3705	8160	4.8	189	1.1	43
Pratt & Whitney	F119-100	157.5	35										
Rolls-Royce	Viper 680-43	–	–	19.4	4.36	0.98	0	380	836	1.96	77	0.73	29
Rolls-Royce	Pegasus 11-61	–	–	106	23.8		1.2	1934	4260	3.48	137	1.22	48
Rolls-Royce/Turbomeca	Adour871	–	–	26.6	5.99	0.74	0.8	590	1300	1.95	77	0.56	22
Snecma	M53-P2	95	21.4	65	14.6	0.9 (dry)		1500	3307	5.0	197	1.0	39
Snecma	M88-2	72.9	16.4	48.7	10.95	0.78 (dry)				3.54	140		
Turbo-Union	RB199–104	73	16.4	40.5	9.1	1.76 (wet)	1.1 (approx.)	976	2150	3.6	142	0.72	28

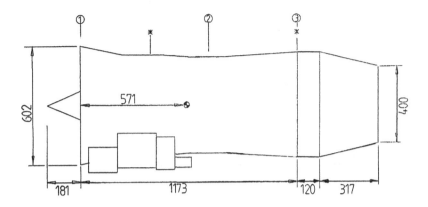

Fig. A5.1 T-91 Project engine dimensions.

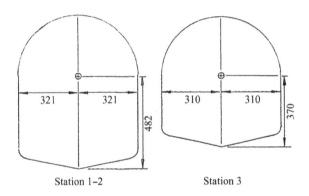

Station 1–2   Station 3

Fig. A5.2 Engine envelope.

dimensions, allowances must be made for accessories, piping, wiring etc. Figure A5.2 shows notional airframe clearance envelopes at several stations along the engine. The sections shown can be assumed to change smoothly from station 2 to station 3. No allowance is made for thermal insulation on the engine outer case, the maximum temperature of which would be expected to be ~464 K.

Figure A5.3 shows the engine installed in the fuselage of the T-91 aircraft.

A5.2.1.2 Engine Data

Total engine mass, less mountings	400 kg
Maximum thrust, sea level static ISA conditions	28.7 kN
Bypass ratio	1.5:1

Throttle response:

(i) On ground, from idle to 95% max thrust	8.5 s
(ii) In flight, from idle to 95% max thrust	5.5 s

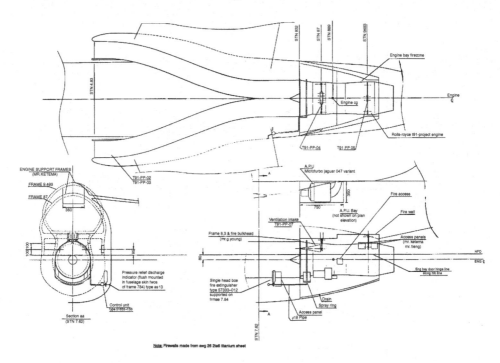

Fig. A5.3 T-91 engine installation.

Polar moments of inertia:

(i)	HP shaft (rotates clockwise as seen from front)	0.18 kg m^2
(ii)	LP shaft (rotates anti-clockwise as seen from front)	1.04 kg m^2
(iii)	Static engine components about shaft axis	14.6 kg m^2
(iv)	Total engine about transverse axes through CG (both equal)	46.4 kg m^2

CG location (on shaft axis), from compressor inlet 0.571 m
Spool angular speeds are given in Table A5.4.

A5.2.1.3 Engine Performance
Engine performance, uninstalled, is shown in Table A5.5. It includes representative bleed and power off-takes for the T-91 application.

Figures A5.4 and A5.5 give plots of the installed engine performance predictions.

A5.2.2 The TF-89 Project Fighter Turbo-Fan Engine
This engine was designed by an engine company's project office to support the Cranfield TF-89 advanced tactical fighter project [47]. Manufacturers information was augmented by runs of the Cranfield TURBOMATCH engine design program.

Table A5.4   Spool angular speeds

Ratings	LP spool ω_{LP}		HP spool ω_{HP}	
	% (rpm)	rad/s	% (rpm)	rad/s
100% rpm (assume max continuous)	100(15 780)	1652.478	100(32 220)	3374.070
Max war rating	105(16 569)	1735.102	105(33 831)	3542.773

Table A5.5   T-91 Turbo-fan project engine performance at sea level

XM	W1A	FN	SFC	WFT	WB3	PWXH
0.000	48.10	28.7	17.969	516.1	0.000	0
0.000	48.10	28.7	17.970	516.1	0.000	0
0.000	49.54	27.7	17.453	483.4	0.360	35
0.000	49.54	27.7	17.453	483.4	0.360	35
0.000	43.67	20.8	15.495	331.2	0.360	35
0.000	36.40	13.8	15.539	215.2	0.360	35
0.000	26.75	6.9	15.974	110.6	0.360	35
0.200	50.74	25.4	19.684	499.2	0.360	35
0.200	50.74	25.4	19.684	499.2	0.360	35
0.200	45.09	19.0	18.391	349.8	0.360	35
0.200	38.12	12.7	18.391	233.0	0.360	35
0.200	29.19	6.3	19.994	126.8	0.360	35
0.200	24.15	3.8	22.420	85.3	0.360	35
0.300	52.27	24.9	20.861	519.6	0.360	35
0.300	46.60	18.7	19.653	367.2	0.360	35
0.300	39.57	12.5	19.809	246.7	0.360	35
0.300	30.96	6.2	21.975	136.8	0.360	35
0.300	26.13	3.7	25.153	94.0	0.360	35
0.500	57.37	25.3	23.290	589.5	0.360	35
0.500	51.36	19.0	22.180	421.1	0.360	35
0.500	43.83	12.7	22.561	285.6	0.360	35
0.500	35.46	6.3	25.656	162.4	0.360	35
0.500	31.03	3.8	30.243	114.8	0.360	35
0.600	61.06	26.2	24.536	641.7	0.360	35
0.600	54.74	19.6	23.438	459.7	0.360	35
0.600	46.75	13.1	23.868	312.1	0.360	35
0.600	38.16	6.5	27.262	178.3	0.360	35
0.600	33.89	3.9	32.442	127.3	0.360	35

Table A5.5   *(cont.)*

XM	W1A	FN	SFC	WFT	WB3	PWXH
0.700	65.61	27.4	25.813	707.5	0.360	35
0.700	58.87	20.6	24.708	507.9	0.360	35
0.700	50.29	13.7	25.151	344.7	0.360	35
0.700	1.11	6.9	28.669	196.5	0.360	35
0.700	36.99	4.1	34.282	141.0	0.360	35
0.800	70.07	27.7	26.734	740.4	0.360	35
0.800	62.65	20.8	25.924	538.5	0.360	35
0.800	53.53	13.8	26.593	368.3	0.360	35
0.800	44.01	6.9	30.344	210.1	0.360	35
0.900	74.85	27.9	27.750	773.5	0.360	35
0.900	66.76	20.9	27.211	568.9	0.360	35
0.900	57.13	13.9	28.093	391.5	0.360	35
0.900	47.29	7.0	32.123	223.8	0.360	35

XM = Mach number; WIA = airflow, (kg/s); FN = nett thrust (kN); SFC = specific fuel consumption, (g/s/kN); WFT = fuel flow (g/s); WB3 = customer bleed (kg/s); PWXH = HP power offtake, (kW).

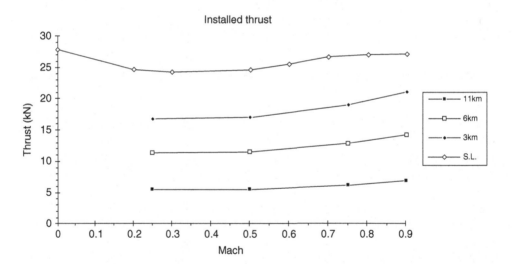

Fig. A5.4 T-91 installed maximum thrust.

A5.2.2.1 Engine Description

The engine has two shafts and a bypass ratio of 0.4. The engine parameters are:

USAF standard intake

Fan pressure ratio   4.5

Compressor pressure ratio   5.5

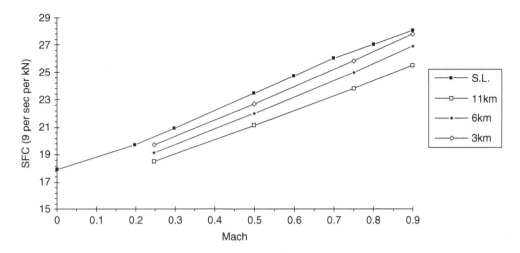

Fig. A5.5 T-91 fuel consumption at maximum thrust.

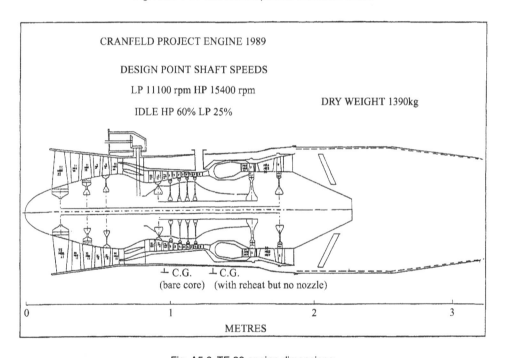

Fig. A5.6 TF-89 engine dimensions.

Fan duct relative total pressure losses:	4.5%
Cooling bleed	1%
Burner combustion efficiency	0.995
Relative total pressure losses	4.4%
Turbine entry total temperature (TET)	1845 K

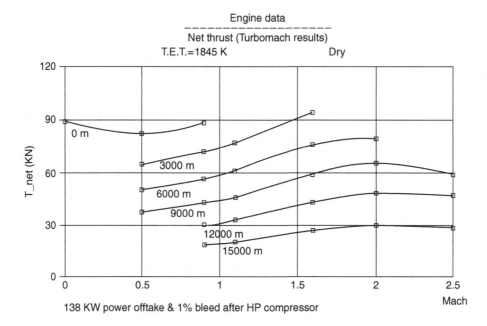

Fig. A5.7 TF-89 net thrust – dry.

Afterburner combustion efficiency	0.99
Relative total pressure losses	2.5%
Maximum total temperature	2300 K
'Condiv' nozzle	
Maximum continuous net thrust	
at M 1.6, 50 000 ft (design point)	28.7 kN

Figure A5.6 shows the engine general arrangement with associated masses and centres of gravity.

A5.2.2.2 Performance

Airframe systems required power off-takes of 139 kW and 1% air bleed for each of the two engines. Performance figures (installed) were calculated with the following intake pressure ratios (P_T):

P_r = 0.90 Mach number between 0 and 0.9
P_r = 0.89 M 1.1
P_r = 0.88 M 1.6
P_r = 0.80 M 2.0
P_r = 0.68 M 2.5

Figures A5.7 and A5.8 show net thrust 'versus' Mach number and altitude for dry and reheat engines, respectively. Figures A5.9 and A5.10 give corresponding fuel flows, whilst Figure A5.11 shows re-heat specific fuel consumption.

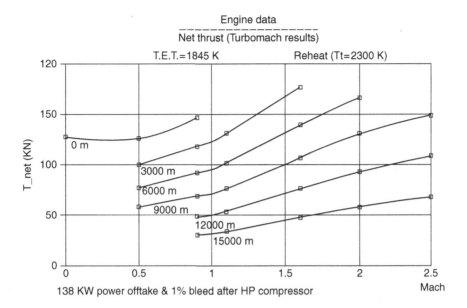

Fig. A5.8 TF-89 net thrust – reheat.

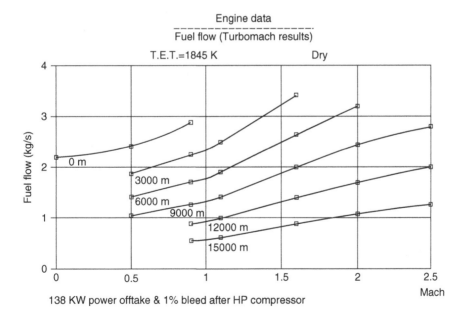

Fig. A5.9 TF-89 fuel flow – dry.

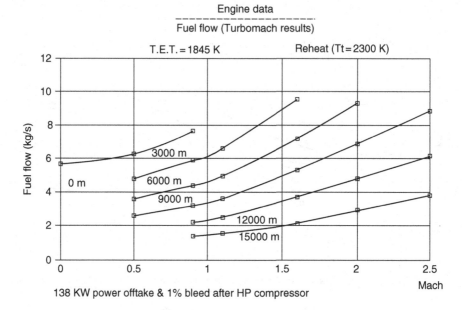

Fig. A5.10 TF-89 fuel flow – reheat.

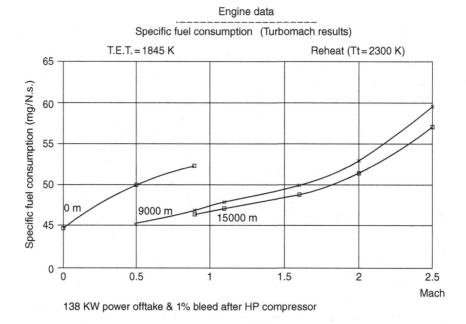

Fig. A5.11 TF-89 specific fuel consumption – reheat.

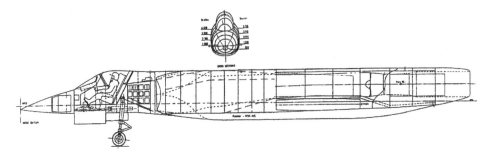

Fig. A5.12 TF-89 fuselage.

A5.2.2.3 Installation

The engine dry mass is 1390 kg, as shown on Fig. A5.6 and its estimated polar moment of inertia is 8.6 kg m^2. Figure A5.12 shows the engine installed in the rear fuselage of the Cranfield TF-89 group project aircraft design.

A5.2.3 The SL-86 Project Turbo-Ramjet Engine

This engine was designed by an engine company's project office to support the Cranfield SL-86 space launcher design project. The latter was a two-stage to orbit horizontal take-off space launcher [48]. The lower stage was a large Concorde-shaped vehicle powered by four of the turbo-ramjets. They were fuelled by liquid hydrogen and operated in turbo-jet mode up to approximately M 2, and then changed to ramjet mode up to the separation Mach number of 4.0. Such engines, however, are suitable for Mach numbers up to 6.0 or 7.0.

The upper stage was a small Space Shuttle type vehicle sitting on top of the launcher until separation. The upper stage was powered by a down-scaled version of the Space Shuttle main engine, fuelled by liquid hydrogen and liquid oxygen.

A5.2.3.1 Engine Description

Type: turbo-ramjet

Sea level static thrust	220 kN
Mass including nozzle, approx.	3500 kg

Further details are shown in Fig. A5.13.

In turbojet mode the engine core takes most of the airflow delivered by the intake, and compresses it through four stages in the LP compressor, and through four stages in the HP compressor. The combustion chamber is similar to a conventional engine, apart from the absence of a flame holder, which is not needed due to the nature of the hydrogen flame, being impossible to 'blow out'. From there the air is expanded through an HP and an LP turbine. The jetpipe houses the afterburner rings, which again do not need a flame holder. The air then passes through a primary nozzle before exiting through the secondary nozzle. A small amount of air passes through the ramjet annulus in this mode, its quantity being governed by the primary nozzle positioning.

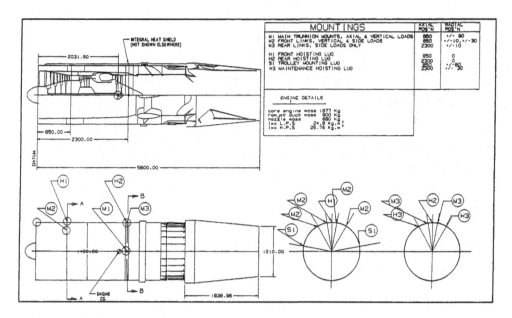

Fig. A5.13 SL-86 turbo-ramjet engine.

In ramjet mode the core still continues to take air, but produces very little thrust, and most air is fed around the ramjet annulus. At the rear of the duct is a row of vanes which occupy one-third of the annulus area. In turbo-jet mode the air flowing through the annulus passes through these vanes and continues through the outer annulus to be mixed with the core air just before the secondary nozzle. The vanes are in two parts, front and rear. The rear half of the vane is rotated about the engine's centerline until it blocks a further one-third of the annulus area; the last third of the area is blocked by a plate which is rotated in the opposite direction and which is normally housed between the two halves of the vanes. At the same time, an area is opened up in the afterburner wall to allow the air into the afterburner chamber, where the ramjet combustion takes place using the same afterburner rings as the turbo-jet mode.

A5.2.3.2 Performance
Figure A5.14 shows a typical SL-86 ascent trajectory and Fig. A5.15 shows the engine performance map. Figure A5.16 shows the liquid hydrogen fuel flow.

A5.2.4 The A-90 Project Airliner Turbo-Fan Engine
The project engine was designed for use on the A-90 500-seat short-range twin-engined airliner [30]. The engine was scaled from a high bypass engine, and is described van den Berg [49].

A5.2.4.1 Engine Description
Figure A5.17 shows a general view of the engine, which is a three-shaft turbo-fan engine with a bypass ratio of about 6.

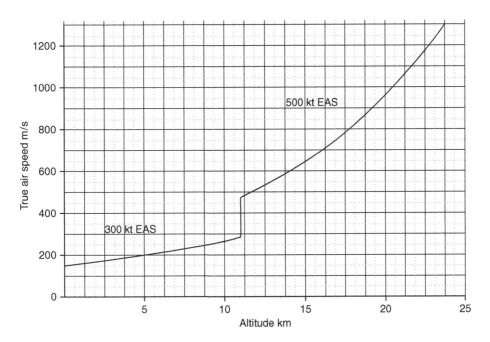

Fig. A5.14 SL-86 ascent trajectory.

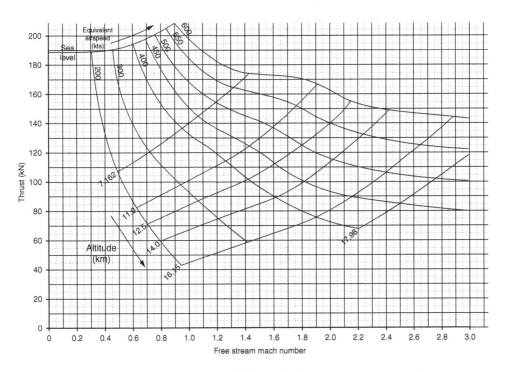

Fig. A5.15 Turbo-ramjet thrust.

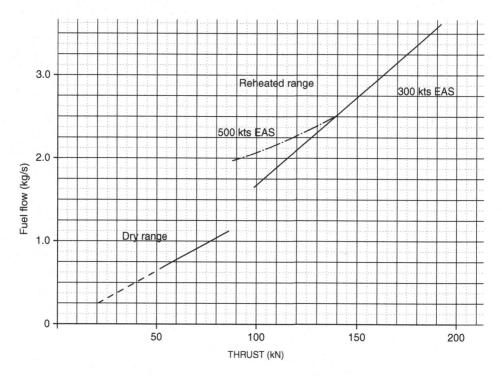

Fig. A5.16 Turbo-ramjet fuel flow.

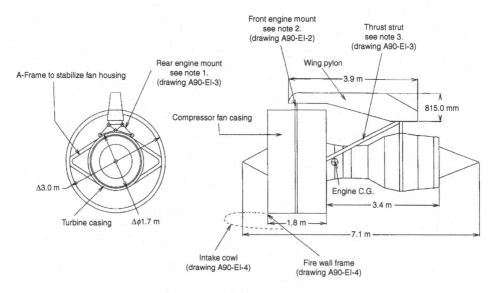

Fig. A5.17 A-90 turbo-fan engine.

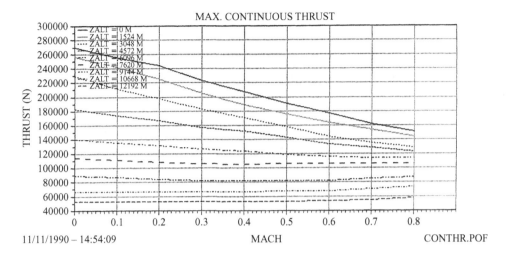

Fig. A5.18 A-90 thrust at maximum continuous rating.

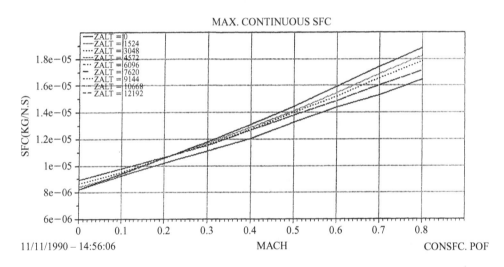

Fig. A5.19 A-90 specific fuel consumption.

Type: high bypass ratio turbo-fan

| Sea level static thrust | 333.6 kN |
| Typical engine mass, approx. | 5000 kg |

A5.2.4.2 Engine Performance

Figure A5.18 shows engine thrust for various Mach numbers and altitudes for the maximum continuous rating and Fig. A5.19 shows the corresponding specific fuel consumption (sfc).

A6 AERODYNAMIC DATA

This section is a compilation of data that should be useful during the conceptual stage of aircraft design. It has been extracted from a number of references and sources available to the author.

A6.1 Basic Aerodynamic Prediction Methods

A6.1.1 The Drag Polar

The drag polar gives a good approximation of aircraft drag for many flight conditions, but can be incorrect at high and low lift conditions.

$$C_D = C_{D_0} + \frac{K C_L^2}{\pi A}$$

where C_{D_0} = zero lift drag coefficient, which is a function of the shape of the aircraft and aircraft Mach number at high subsonic speeds and above. The references in Chapter 9 may be used to estimate this coefficient. The chart in Fig. A6.2 gives a very simple method for use with subsonic jet transports.

C_L = lift coefficient.

K = the vortex drag factor, which depends on the way the lift is distributed across the span. The minimum value of $K = 1.0$ occurs when the spanwise lift distribution is elliptic in form. However, the most important influence on vortex drag is that of aspect ratio A, which should be large to minimize the vortex drag.

Torenbeek [11] and the AIAA [50] imply the following guidelines for K for a number of types of aircraft:

High subsonic jet transport = 1.33–1.18
Large turbo-prop transports = 1.25–1.18
Twin piston-engine aircraft = 1.33–1.25
Single piston-engine
Rectractable gear = 1.33–1.25
Fixed gear = 1.54–1.33

The lower values are for tapered wings, close to elliptic planform.

Additional values are required to predict the drag of landing gears, flaps, weapons, etc. and are available in many text books.

A6.1.2 Flight Performance Formulae and Methods

The following formulae may be used for initial design purposes.

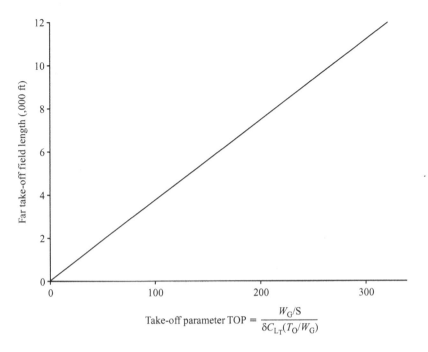

Fig. A6.1 Simple take-off estimation method, d = density ratio; W_G = gross take-off weight (lb); S = gross wing area (ft^2); l_o = sea level static thrust of all engines (lbf); C_{L_T} = Max lift coefficient in take-off configuration (*not* C_L at take-off speed).

A6.1.2.1 Take-Off

Take-off-distance calculations treat ground roll and the distance to clear an obstacle. Obstacle requirements differ for commercial (35 ft) and military (50 ft) aircraft. Take-off ground roll:

$$S_{gnd} = -\frac{W/S}{g\rho(C_D - \mu C_L)} \ln\left[1 - \frac{A_T^2(C_D - \mu C_L)}{((T/W) - \mu)C_{L_{max}}}\right].$$

The stall margin (A_T) typically is 1.2.

Total take-off distance: $S_{TO} = (S_{gnd})(F_{PL})$, where g = acceleration due to gravity; ρ = air density; μ = rolling friction (typically 0.03); T = thrust; W = weight; S = wing area; F_{PL} = the factor to clear an obstacle. It depends greatly upon available excess thrust, flight path and pilot technique. The following typical factors characterize planforms in their ability to clear a 50-ft obstacle: straight wing = 1.15; swept wing = 1.36; delta wing = 1.58.

Figure A6.1 shows an empirical curve to approximately predict take-off field lengths for civil jet transports.

A6.1.2.2 Climb

For small angles, the rate of climb can be determined from

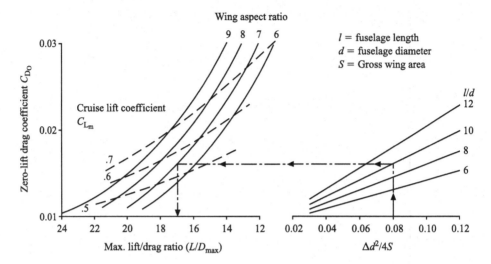

Fig. A6.2 Cruise lift/drag parameters. (Based on Reynold's number for Boeing 747- class aircraft (see [60] Appendix 2) for example of Reynold's number correction.)

$$R/C = (T - D)V/W\left(1 + \frac{V\mathrm{d}V}{g\,\mathrm{d}h}\right).$$

where $V/g \times \mathrm{d}V/\mathrm{d}h$ is the correction term for flight acceleration.

For low subsonic climb speeds, the acceleration term is usually neglected:

$$R/C = (F - D)V/W.$$

$$\gamma = \sin^{-1}\left(\frac{T - D}{W}\right),$$

where V = velocity; D = drag; h = altitude.

A6.1.2.3 Cruise

The basic cruise distance can be determined by using the Breguet range equation for jet aircraft, as follows:

Cruise range: $R = L/D(V/SFC)\ln(W_0/W_1)$ where the subscripts '0' and '1' stand for initial and final weight respectively, and SFC = specific fuel consumption.

Cruise fuel: Fuel $= W_0 - W_1 = W_f(e^{R/k} - 1)$, where k, the range constant, equals L/D (V/SFC).

For large variations in weight, speed and altitude during cruise, it is suggested that the range calculations be divided into increments and summed up for the total.

Fig. A6.2 shows a simplified method to predict cruise drag for civil jet transports, which was extracted from Loftin [60].

A6.1.2.4 Manoeuvres

Measures of manoeuvrability of a vehicle are often expressed in terms of sustained and instantaneous performance. For sustained performance, thrust available from the engines must equal the drag of the vehicle (i.e. specific excess power equals zero).

Instantaneous turns are lift-limited and are also referred to as attained turns.

$$SEP = p_s \left(\frac{T - D}{W} \right) V.$$

Turn radius: $TR = \frac{V^2}{g\sqrt{n^2-1}}.$

Turn rate: $\dot{\psi} = \frac{v}{TR} = \frac{g}{v}\sqrt{n^2 - 1},$

where SEP = specific excess power (p_s in US notation); n = load factor, i.e. manoeuvre acceleration factor. Typically 2.5 g for large transports and 9 g for fighters.

A6.1.2.5 Loiter

Loiter performance is based upon conditions at $(L/D)_{max}$, since maximum endurance is of primary concern.

$$\text{Loiter speed}: M = \sqrt{\frac{W/S}{0.7(\rho)(C_L)_{(L/D)_{max}}}}$$

where M = Mach number.

$$(C_L)_{(L/D)_{max}} = \sqrt{\frac{C_{D_{min}}}{k}}$$

where ρ = atmospheric pressure.

$$\text{Loiter time}: t = L/D_{max} \left(\frac{1}{SFC} \right) \ln \left(\frac{W_0}{W_1} \right).$$

$$\text{Loiter fuel}: Fuel = W_1 e^y \text{ where } y = \left\{ \frac{t(SFC)}{L/D_{max} - 1} \right\}.$$

A6.2.1.6 Landing

Landing distance calculations cover distance from obstacle height to touchdown and ground roll from touchdown to a complete stop.

$$\text{Approach distance}: S_{air} = \left(\frac{V_{obs}^2 - V_{TD}^2}{2g} + h_{obs} \right)(L/D),$$

where V_{obs} = speed at obstacle; V_{TD} = speed at touchdown; h_{obs} = obstacle height; L/D = lift-to-drag ratio (with appropriate high-lift devices).

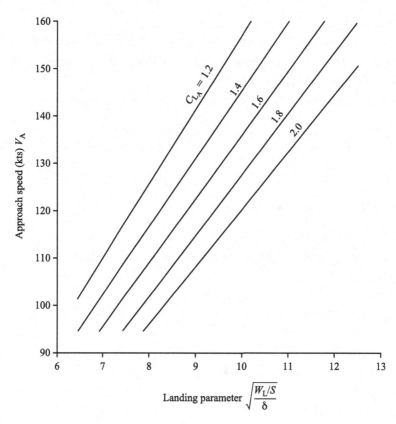

Fig. A6.3 Simple landing parameter. W_L = max landing mass; C_{L_A} = approach lift coefficient = $C_{L_{max}}/1.69$; $C_{L_{max}}$ = maximum lift coefficient in landing configuration. (Note for most aircraft $W_G > W_L$.)

$$\text{Landing ground-roll}: S_{\text{gnd}} = \frac{(W/S)}{g\rho(C_D - \mu_{\text{BRK}}C_L)} \ln\left[1 - \frac{A_L^2(C_D - \mu_{\text{BRK}}C_L)}{((T/W) - \mu_{\text{BRK}})C_{L_{\text{max}}}}\right],$$

where μ_{BRK} = coefficient of braking friction; A_L = the stall margin on landing (typically 1.3).

Normally the distance would require a 2 s delay to cover the time required to achieve full braking. Commercial requirements also dictate conservative factors be applied to the calculated distances.

Figure A6.3 shows an empirical curve to approximately predict landing wing loading for civil jet transports (see also Appendix B).

A6.2 Wing Section and Lift Data

Tables A6.1 and A6.2 give data for low-speed aerofoil section aerodynamic properties and high-speed advanced section aerodynamic properties, respectively.

Table A6.1   Low-speed aerofoil section aerodynamic properties – NACA experimental data from
Abbott and Von Doenhoff [51]

Aerofoil	α_0 (deg)	C_{M0}	a_{1rad}	$C_{l_{\alpha/deg}}$	$\alpha_{C_{L_{max}}}$ (deg)	$C_{L_{max}}$	Section C_d at $C_{L_{max}}$
0006	0	0	6.19	0.108	9.0	0.92	0.0095
0009	0	0	6.25	0.109	13.4	1.32	0.0124
23012	−1.4	−0.014	6.13	0.107	18.0	1.79	0.016
23015	−1.0	−0.007	6.13	0.107	18.0	1.72	0.02
23018	−1.2	−0.005	5.96	0.104	16.0	1.60	0.016
23021	−1.2	0	5.90	0.103	15.0	1.50	0.0162
63–006	0	0.005	6.42	0.112	10.0	0.87	0.0086
63–009	0	0	6.36	0.111	11.0	1.15	0.0113
63–206	−1.9	−0.037	6.42	0.112	10.5	1.06	0.008
63–209	−1.4	−0.032	6.30	0.11	12.0	1.4	0.0127
63–210	−1.2	−0.035	6.47	0.113	14.5	1.56	0.014
63_1–012	0	0	6.65	0.116	14.0	1.45	0.0134
63_1–212	−2.0	−0.035	6.53	0.114	14.5	1.63	0.0117
63_1–412	−2.8	−0.075	6.70	0.117	15.0	1.77	0.0154
63A 010	0	0.005	6.02	0.105	13.0	1.2	0.0146
63A 210	−1.5	−0.04	5.9	0.103	14.0	1.43	0.014
64–006	0	0	6.25	0.109	9.0	0.8	0.007
64–009	0	0	6.3	0.11	11.0	1.17	0.0126
64–206	−1.0	−0.04	6.3	0.11	12.0	1.03	0.009
64–210	−1.6	−0.04	6.3	0.11	14.0	1.45	0.0118
64_1–412	−2.6	−0.065	6.42	0.112	15.0	1.67	–
64A 010	0	0	6.3	0.11	12.0	1.23	0.011
64A 210	−1.5	−0.04	6.02	0.105	13.0	1.44	0.011
64A 410	−3.0	−0.08	5.73	0.10	15.0	1.61	0.012
64_1A 212	−2.0	−0.04	5.73	0.10	14.0	1.54	0.012
64_2A 215	−2.0	−0.04	5.44	0.095	15.0	1.5	0.016
65–006	0	0	6.02	0.105	12.0	0.92	0.008
65–009	0	0	6.13	0.107	11.0	1.08	0.012
65–206	−1.6	−0.031	6.02	0.105	12.0	1.03	0.009
65–210	−1.6	−0.034	6.19	0.108	13.0	1.4	0.0137

$R_e = 9 \times 10^6$, smooth leading-edge.

The section information notation is a_0 = zero-lift angle of attack; α_1 = section lift-curve slope (British); C_{l_α} = section lift-curve slope (US); $C_{L_{max}}$ = section maximum lift coefficient; and C_{M_0} = section zero-lift pitching moment coefficient.

The NACA numbering system for four-digit series is shown in Table A6.1.

Table A6.2   Section aerodynamic properties: advanced aerofoil sections at high subsonic speeds

Aerofoil	C_L	C_{Mo} at cruise	a_1 (rad)	$C_{L\alpha}$ (deg)	$c_{l_{max}}$ (Low speed)	Section C_d at given cruise C_L at M_D	t/c max	M drag rise at given C_L	Ref for data
RAE9515	0.3	−0.106	6.47	0.163	1.0	0.013	0.105	0.79	RAE R&M 3820, 1978 [52]
	0.4	−0.106	(Low speed)			0.013		0.79	
	0.5	−0.11				0.013		0.79	
	0.6	−0.12				0.014		0.79	
RAE9530	0.3	−0.12	5.84	0.102	1.23	–	0.105	0.80	RAE R&M 3820, 1978 [52]
	0.4	−0.12	(Low speed)			0.015		0.795	
	0.5	−0.13				0.015		0.79	
	0.6	−0.5				0.018		0.78	
RAE9550	0.3	−0.08	6.88	0.12	1.08		0.122	0.775	RAE R&M 3820, 1978 [52]
	0.4	−0.09						0.77	
	0.5	0.095						0.765	
	0.6	−0.01						0.72	
RAE5225	0.3	−0.08			–	0.0107	0.14	–	RAE TR 87002 [53]
	0.4	−0.10			all at M = 0.735	0.0107		–	
	0.5	−0.1				0.011		–	
	0.6	−0.1				0.0114		0.735	
RAE5230	0.3	−0.1			all at M = 0.735	0.0112	0.14	–	RAE TR87002 [53]
	0.4	−0.1				0.0111		–	
	0.5	−0.1				0.0114		–	

RAE5236	0.5	-0.1		0.012		0.735	RAE TR87002 [53]
	0.3	-0.076		0.0111	0.14		–
	0.4	-0.078	all at	0.0113			–
	0.5	-0.076	M = 0.735	0.0116			
	0.6	-0.076		0.0122		0.735*	
NACA SC(3)	0.3	-0.17	all at	0.0122	0.12		NASA TM–86371 [54]
			M = 0.78				–
-0712(B)	0.4	-0.18	$R_e = 30$	0.014			–
	0.5	-0.18	$\times 10^6$	0.018			
	0.6	-0.18		0.021			
DSMA523	0.3	–	all at	0.012	0.11	0.80**	NASA TM–81336 [55]
	0.4		M = 0.8	0.012			
	0.5		$R_e = 14.5 \times 10^6$	0.013			
	0.6			0.016		0.8	

* At C_L = 0.58
** at C_L = 0.68.

The first integer indicates the maximum value of the mean-line ordinate y_c in per cent of the chord. The second integer indicates the distance from the leading edge to the location of the maximum camber in tenths of the chord. The last two integers indicate the section thickness in per cent of the chord. Thus the NACA 2415 wing section has 2% camber at 0.4 of the chord from the leading edge and is 15% thick.

The system for the five-digit series is:

The first integer indicates the amount of camber in terms of the relative magnitude of the design lift coefficient; the design lift coefficient in tenths is thus three-halves of the first integer. The second and third integers together indicate the distance from the leading edge to the location of the maximum camber; this distance in per cent of the chord is one-half the number represented by these integers. The last two integers indicate the section thickness in per cent of the chord. The NACA 23012 wing section thus has a design lift coefficient of 0.3, has its maximum camber at 15 per cent of the chord, and has a thickness ratio of 12 per cent.

The system for the six series means:

The first digit, 6, identifies the aerofoil as a member of the 6-series low drag family. The second digit is the chordwise position of the minimum pressure in tenths of the chord behind the leading edge for the basic symmetric section at zero lift. The figure written in subscript gives the range of lift coefficient in tenths above and below the design lift coefficient in which pressure gradients favourable for obtaining low drag exist on both surfaces. The digit following the subscript gives the design lift coefficient in tenths. The last two digits give the thickness of the wing in percentage of the local chord.

A7 STRUCTURES AND MATERIALS DATA

A7.1 Basic Structural Formulae

Load transfer methods are described in the main text of this book, in Section 5.2.2 and the associated simple formulae are as follows.

A7.1.1 Tension

$$f_t = \frac{P}{A},$$

where f_t is the stress (tensile in this case), P is the applied load (tensile), and A is the cross-sectional area of the component.

A7.1.2 Compression

(a) Pure compression (without buckling) $f_c = \frac{P}{A}$,

where f_c is the compressive stress and P is the applied load (compressive).

(b) Overall buckling

Unless the strut is short the overall buckling can be estimated by using the Euler formula:

$$P_{\text{crit}} = \frac{\pi^2 EI}{l^2},$$

where E is the elastic modulus, I the second moment of area and P_{crit} the failing load. The value of the length, l, is chosen as the equivalent pin-ended length and although it is obvious in some cases, it is often a matter requiring considerable experience.

(c) Skin compression buckling

Many aircraft components include or consist of flat, or nearly flat panels, and are often subjected to compression. The instability stress in such a case is defined as:

$$f_{\text{crit}} = KE\left(\frac{t}{b}\right)^2$$

where t is thickness, b is length of the smallest side and K is a function of the 'edge fixation', and the ratio of the lengths of the sides. $K = 3.62$ for simply supported, i.e. pinned edges and 6.31 for clamped edges. It is usually reckoned that structural members supporting a panel with a single row of attachments provide simple support conditions, whereas a multiple row of attachments provides clamped conditions, as an assumption for initial design purposes. In the event of one edge being pinned and the parallel edge being free, $K = 0.58$.

Curved panels occur frequently in aircraft structures and the basic buckling stress in this case is given by:

$$f_{\text{buckling}} = 0.6 \frac{Et}{R},$$

where R is the radius of curvature. The factor 0.6 is not attained in practice, and 0.4 is probably more typical. The addition of internal pressure increases the buckling stress value.

A7.1.3 Shear
Skin shear flow $q = \frac{P_s}{l}$,

where P_s is the shear load and l is the length of the shear panel subject to the load shear stress, $f_s = q/t = P_s/A$, where t is the skin thickness and $A = l \times t$ = shear area.

A 7.1.4 Bending
The well known 'engineers' beam theory is adequate for many applications although it may be in error at the ends of a beam where so called 'constraint effects' become important.

$$\frac{f}{y} = \frac{M}{I},$$

where f is the stress, M the applied moment and y the distance from the plane of zero strain.

A7.1.5 Torsion

The well known relationship for a *solid* torsion member is:

$$\frac{G\theta}{l} = \frac{F_s}{r} = \frac{T}{J}$$

and this is quite adequate for most applications. J is the polar second moment of area, G is the shear modulus, r the cross-section radius, T the applied torque f_s the shear stress, and θ is the total angular deflection over the length, l.

When a *thin-walled* section is being considered the 'Batho' formula can be used:

$$q = \frac{T}{2A},$$

where q is the shear flow and A is the enclosed area.

The corresponding formula for twist is:

$$\frac{\theta}{l} = \frac{T}{4GA}\sum\frac{P_e}{t},$$

where P_e is the perimeter and t is the wall thickness.

A torsion member, particularly one of thin-walled section, will be subjected to constraint effects at a built-up end, just as a beam is.

A7.1.6 Pressure Vessels

Pressure vessels often occur in aeronautical applications, for example in cabins and fuel tanks. The loads resulting from internal pressure are tension in character. The basic pressure on the walls of a cylinder gives a hoop tension:

$$f_h = \frac{PR}{t},$$

where p is the pressure.

The pressure load on the end caps gives a longitudinal tension normal to the hoop tension of:

$$f_h = \frac{PR}{2t}.$$

A8 Landing-Gear Data

Section 6.5 of Chapter 6 of this book gives some background on various landing-gear configurations, layout guidelines and landing-gear illustrations. The contact between the aircraft and the ground has consequences in terms of damage to runways or taxiways. An aircraft's performance in this respect is termed its flotation characteristics. The obsolete measure used for commercial aircraft was termed its load classification number (LCN). This term has now been replaced by aircraft classification number (ACN). Figure A8.1 shows typical figures for current transports and a number of options considered for the Cranfield A-94 600 passenger long-range airliner [56].

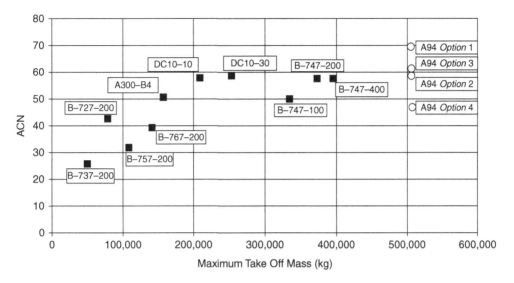

Fig. A8.1 Main gear – flotation.

Other vital data required by aircraft designers are the dimensions and capabilities of currently available aircraft wheels and tyres. Table A8.1 gives such data for type III lower-pressure tyres. Table A8.2 gives similar information for higher-pressure type VII tyres.

A9 AIRCRAFT INTERIOR DATA

This appendix assembles the information that is vital for realistic fuselage interior design for both civil and military aircraft. It starts with the dimensions of human beings, which are important whether they be crew or passengers. The human theme is continued to consider flight-deck and cockpit arrangements and then passenger seating and amenities. This theme is then discontinued and followed by information about standard cargo containers and avionic box dimensions.

A9.1 Human Dimensions

Human beings come in many shapes and sizes, and there are many anthropomorphic standards. Figure A9.1 gives a readily usable list of dimensions for extremes of size percentages of a population of male crew members [58]. Other tables are available for other populations, and many of them have been incorporated into articulated computer-aided design models.

A9.2 Flight-Deck Arrangements

The flight-deck is the vital work-place of the flight crew and must be designed for the maximum safety and efficiency. Figure A9.2 shows the arrangement for a narrow-body flight-deck together with its crew vision envelope. This aircraft was the project which led to the design of the Airbus A320.

Table A8.1   Dimensions and capabilities of lower pressure tyres (Type III)

Type	Nominal diameter (in)	(cm)	Nominal width (in)	(cm)	Rated load (static) (lb)	(kg)	Maximum rated speed (mph)	(km/h)	Normal loaded radius (in)	(cm)	Tyre weight (lb)	(kg)	Inflation pressure (lb/in²)	(mb)
3.00–3.5	8.5	21.6	3.0	7.6	410	186	–	–	3.6	9.1	1.76	0.8	50	3448
4.00–3.5	10.75	27.3	4.0	10.2	710	322	160	257	4.45	11.3	3.1	1.4	40	2758
4.00–4	11.0	27.9	4.0	10.2	940	427	160	257	4.75	12.1	4.0	1.8	70	4827
5.00–4.5	13.0	33	5.0	12.7	–	–	160	257	3.9	9.9	7.5	3.4	78	5378
5.00–5	14.0	35.6	5.0	12.7	800–3100	363–1407	120–160	193–257	5.7	14.5	4.9–7.0	2.2–3.2	31–130	2137–8964
5.5–4	13.25	33.7	5.5	14.0	1225	556	160	257	5.8	14.7	8.0	3.6	50	3448
6.00–4	15.5	39.4	6.0	15.2	1450	658	160	257	6.65	16.9	7.5	3.4	35	2413
6.00–6	17.0	43.2	6.0	15.2	1150	522–1067	120–160	193–257	6.9	17.5	7.6–8.8	3.5–4.0	29–55	2000–3792
6.00–6.5	17.25	43.8	6.0	15.2	2350	795	160	257	7.0	17.8	8.4	3.8	45	3100
6.50–10	21.5	54.6	6.5	16.5	1750	1700	160–190	257–306	9.0	22.9	11.90–17.2	5.4–7.8	160–190	11 030–13 100
7.00–6	18.5	47.0	7.0	17.8	3750	3768	120	193	7.3	18.5	9.0–10.9	4.1–4.9	56–110	3860–7585
7.50–10	23.7	60.2	7.5	19.0	8300	1135	160	257	9.7	24.6	16.8	7.6	63	4344
8.00–7	20.0	50.8	8.0	20.3	2500	2452	161	258	8.06	20.5	16.0	7.3	60	4137
8.50–10	25.0	63.5	8.5	21.6	5400	1762	120–160	193–257	10.2	25.9	21–33	9.5–15.0	41–100	2827–6895
9.00–6	22.0	55.9	9.0	22.9	3880	1521	160	257	8.75	22.2	24	10.9	58	4000
11.00–12	31.5	80.0	11.0	27.9	3350	1476	160	257	8.25	21.0	39.4	17.9	65	4482
12.50–16	38	96.5	12.5	31.8	3250	3632	160	257	15.82	40.2	71.3	32.4	75	5171
20.00–20	55	139.7	20	50.8	3000	2043	200	322	22.4	56.9	251.4	114.1	125	8619
					4500	4011								
					8835	5811								
					12 800	21111								
					46 500									

This is the most popular low-pressure tyre found today on smaller aircraft. The section is relatively wide in relation to the bead. This provides lower pressure for improved cushioning and flotation. The section width and rim diameter are used to designate the size of this tyre. (Information has been extracted from [57].)

Note: steady braked load - 1.5 × rated load and full closure load = 2.7 rated load (approx).

Table A8.2 Dimensions and capabilities of higher pressure tyres (Type VII)

Type	Plys	Nominal outside diameter (in)	(cm)	Nominal width (in)	(cm)	Rated load (static) (lb)	(kg)	Maximum rated speed (mph)	(km/h)	Normal loaded radius (in)	(cm)	Tyre weight (lb)	(kg)	Inflation pressure (lb/in²)	(mb)
13½ × 4.2	6	13.5	34.5	4.25	10.8	2310	1049	160	257	5.75	14.6	4.6	2.1	160	11 030
5		16.0	40.6	4.4	11.2	2300	1044	185	297	6.9	17.5	9.9	4.5	120	8275
16 × 4.4	12	20.0	50.8	4.4	11.2	5150	2336	225	361	8.9	22.6	14.9	6.8	225	15 510
20 × 4.4	12	18.0	45.7	5.5	14.0	5050	2293	190	305	7.5	19.0	11.3	5.1	170	11 720
18 × 5.5	18	18.0	45.7	5.7	14.5	8600	3904	250	402	7.5	19.0	15.7	7.1	300	20 680
18 × 5.7	14	18.0	45.7	5.7	14.5	6200	2815	250	402	7.5	19.0	15.5	7.0	215	14 820
18 × 5.7	8	17.5	44.5	6.25	15.9	2900	1317	190	305	6.9	17.5	13.0	5.9	70	4825
17.5 × 6.2	12	18.0	45.7	6.5	16.5	5000	2270	256	411	7.6	19.3	13.6	6.2	150	10 340
5	16	22.0	55.9	6.5	16.5	10 000	4540	225	361	9.4	23.9	23.3	10.6	250	17 235
18 × 6.5	12	26.0	66.0	6.6	16.8	8600	3904	225	361	11.2	28.4	29.8	13.5	185	12 750
22 × 6.5	20	22.0	55.9	6.6	16.8	12 000	5448	218	350	9.35	23.7	26.2	11.9	270	18 610
26 × 6.6	8	22.0	55.9	6.75	17.1	4450	2020	120	193	9.1	23.1	14.5	6.6	95	6550
22 × 6.6	10	22.0	55.9	6.75	17.1	5900	2679	125	201	9.1	23.1	18.0	8.2	125	8620
22 × 6.75	12	23.0	58.4	7.0	17.8	7800	3541	210	337	9.9	25.1	24.3	11.0	160	11 030
22 × 6.75	10	24.0	61.0	7.25	18.4	6 600	2 996	200	321	10.4	26.4	31.0	14.1	120	8275
23 × 7	14	24.0	61.0	7.7	19.6	8200	3723	190	305	10.0	25.4	29.4	13.3	135	9305
24 × 7.25	16	24.0	61.0	7.7	19.6	9725	4415	210	335	10.0	25.4	29.8	13.5	165	11 375
24 × 7.7	12	27.0	68.6	7.75	19.7	9650	4381	225	361	11.8	30.0	35.4	16.1	200	13 790
24 × 7.7	12	26.5	67.3	8.0	20.3	9500	4313	160	257	11.13	28.3	29.8	13.5	150	10 340
27 × 7.75	20	25.5	64.8	8.0	20.3	15 300	6 946	250	402	10.95	27.8	36.5	16.6	250	17 235
26.5 × 8	12	24.5	62.2	8.5	21.6	7400	3 360	190	305	10.1	25.7	28.9	13.1	110	7240
24.5 × 8.5	12	32.0	81.3	8.8	22.4	11 000	4994	190	305	13.3	33.8	41.3	18.8	140	9650
0	12	28.0	71.1	9.0	22.9	8800	3995	160	257	11.4	29.0	36.1	16.4	100	6895
26 × 8.75	22	36.0	91.4	11.0	27.9	23 300	10 578	190	305	14.7	37.3	95	43.1	190	13 100
32 × 8.8	2612	30.0	76.2	11.5	29.2	32 170	14 605	250	402	12.5	31.8	80	36.3	332	22 890

Table A8.2 (cont.)

Type	Plys	Nominal outside diameter (in)	(cm)	Nominal width (in)	(cm)	Rated load (static) (lb)	(kg)	Maximum rated speed (mph)	(km/h)	Normal loaded radius (in)	(cm)	Tyre weight (lb)	(kg)	Inflation pressure (lb/in^2)	(mb)
28 × 9	12	32.0	81.3	11.5	29.2	11200	5085	210	335	13.58	34.5	60.8	27.6	120	8275
36 × 11	22	39.0	99.1	13.0	33.0	24 600	11 168	210	335	15.6	39.6	104	47.2	165	11 375
30 × 11.5	20	31.0	78.7	13.0	33.0	17 200	7809	225	361	12.7	32.3	71.2	32.3	155	10 685
32 × 11.5	6	36.0	91.4	13.0	33.0	6350	2883	160	257	14.1	35.8	53.2	21.1	35	2410
39 × 13	28	40.0	101.6	14.0	35.6	33 100	15 027	225	361	16.45	41.8	126	57.2	200	13 790
31 × 13	24	37.0	94.0	14.0	35.6	25 000	11 350	225	361	15.1	38.4	105	47.7	160	11 030
36 × 13	20	40.0	101.6	14.0	35.6	27 100	12 303	225	361	16.6	42.2	116	52.7	165	11 375
40 × 14	26	40.0	101.6	14.5	36.8	36 800	16 707	225	361	16.7	42.4	150	68.1	220	15 165
37 × 14	20	42.0	106.7	15.0	38.1	23 500	10 670	190	305	17.3	43.9	109	49.5	120	8275
40 × 14	28	40.0	101.6	15.5	39.4	39 500	17 933	225	361	16.1	40.9	148	67.2	195	13 440
40 × 14.5	32	47.0	119.4	15.75	40.0	51 481	23 372	278	446	20.25	51.4	220	99.9	223	15 375
42 × 15	32	46.0	116.8	16.0	40.6	48 000	21 792	225	361	13.25	33.7	189	85.8	245	16 890
40 × 15.5	28	44.5	113.0	16.5	41.9	42 800	19 430	225	361	18.3	46.5	209	94.9	195	13 440
47 × 15.75	32	49.0	124.5	17.0	43.2	50 400	22 880	210	335	20.2	51.3	223	101.2	210	14 475
46 × 16	32	46.0	116.8	18.0	45.7	51 100	23 200	225	365	19.2	48.8	241	109.4	230	15 855
44.5 × 16.5	34	49.0	124.5	19.0	48.3	55 700	25 290	235	377	20.3	51.6	249	113.0	215	14 820
49 × 17	36	50.0	127.0	20.0	50.8	58 800	26 695	225	365	20.6	52.3	283	128.5	210	14 475
46 × 18	24	56.0	142.2	20.0	50.8	38 500	17480	200	321	22.7	57.7	243	110.3	110	7580
49 × 19	30	52.0	132.1	20.5	52.1	63 700	28 920	235	377	21.3	54.1	299	135.7	195	13 440
50 × 20	36	54.0	137.2	21.0	53.3	72 200	32 780	212	341	22.2	56.4	341	154.8	212	14 615

These extra-high-pressure tyres are the standard for jet aircraft. They have high load-carrying ability and are available in ply ratings from 4 to 38. The higher the number of reinforcing plys, the higher is the pressure that may be used. The tyre sizes are designated by their outside diameter and section width.

		PERCENTILE	
		3rd	99th
G	Shoulder height, sitting	614	727
H	Sitting eye height	765	896
J	Sitting height	876	1007
K	Vertical functional reach, sitting	1281	1515
L	Knee Height, sitting	514	623

		PERCENTILE	
		3rd	99th
A	Bideltoid breadth	427	514
B	Biacromial breadth	370	452
C	Hip breadth, sitting	332	415
D	Stool height	376	471
E	Thigh clearance height	137	191
F	Acromial height, sitting	558	681

		PERCENTILE	
		3rd	99th
M	Functional reach	736	889
N	Cervicale height, sitting	627	745
O	Elbow rest height, sitting	200	306
P	Stomach depth	203	306
Q	Buttock – knee length	558	672
R	Buttock – heel length	998	1211

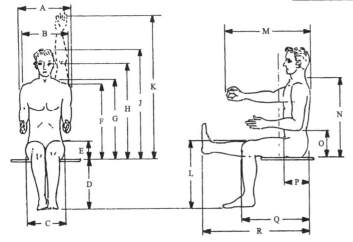

Fig. A9.1 RAF aircrew third and 99th percentile values (mm).

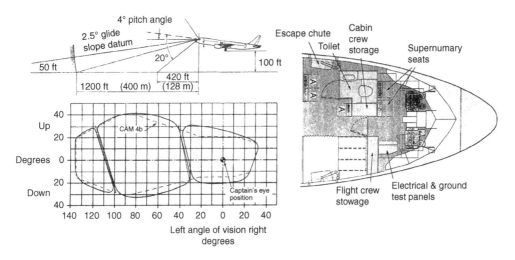

Fig. A9.2 Narrow-body jet flight-deck.

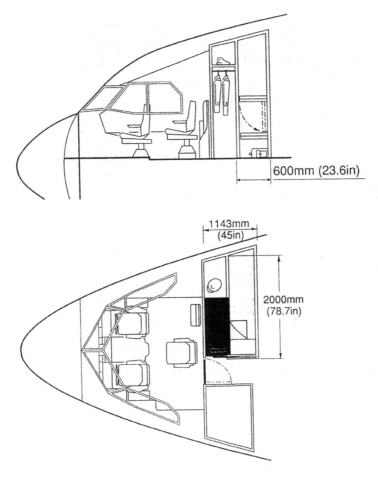

Fig. A9.3 A340 flight deck and crew rest compartment.

Figure A9.3 shows the flight deck of the Airbus A340, together with a flight-crew rest compartment.

A9.3 Cockpit Arrangements

Figure A9.4 shows the definition of the main reference points for military cockpits [58]. Figure A9.5 shows typical cockpit arrangements in the inboard profile of the SAAB Gripen. Chapter 6 shows typical modern flight instrument panels.

A9.4 Passenger Seating

Figure A9.6 shows a cabin plan-view of a typical narrow-body jet transport, the Fokker 100, and its corresponding cross-section is shown in Fig. A9.7. A cabin plan-view of an Airbus A340–200 is shown in Chapter 6 and cross-sections are shown in Fig. A9.8. These are typical of wide-body transports.

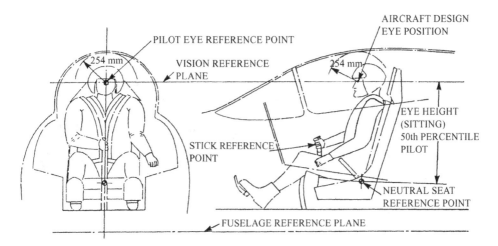

Fig. A9.4 Military cockpit reference points

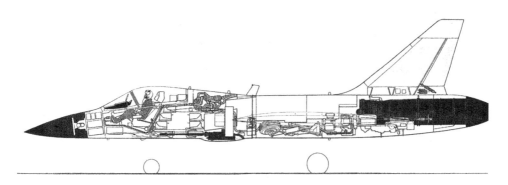

Fig. A9.5 Cutaway of SAAB Gripen 39C

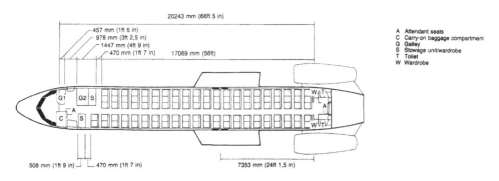

Fig. A9.6 Cabin layout for Fokker 100.

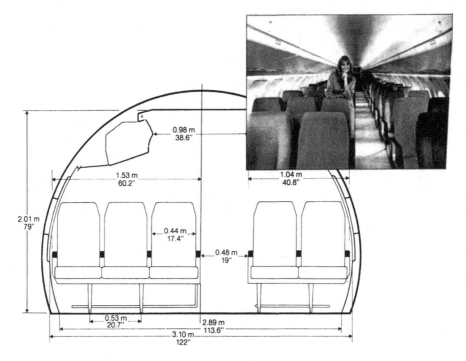

Fig. A9.7 Fokker 100 cross-section.

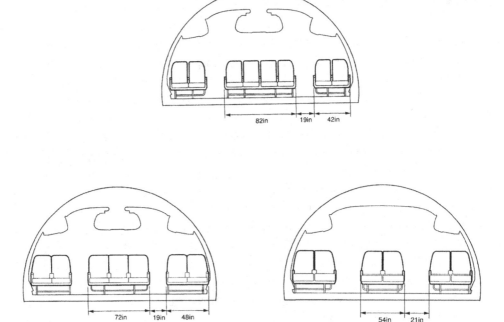

Fig. A9.8 A340 cross-sections. Top, economy class; left, business class; right, first class.

A9.5 Passenger Amenities

The above figures show seating, but also the positions of passenger amenities. These are shown in more detail as:

(i) Toilet compartments for Fokker 100 in Fig. A9.9.
(ii) Galley compartments for Fokker 100 in Fig. A9.10.
(iii) Overhead storage and video for Airbus A330 in Fig. A9.11.

A9.6 Cargo Containers and Pallets

Figure A9.12 shows the range of under-floor cargo containers that may be carried by the wide-body A300/A340 family. The major dimensions of containers is also shown. Figure A9.13 shows the under-floor capacity of the Airbus A320 narrow-body aircraft. The LD3-46 is a low-height container developed from the base of a standard LD3 container. Other five- or six-passenger

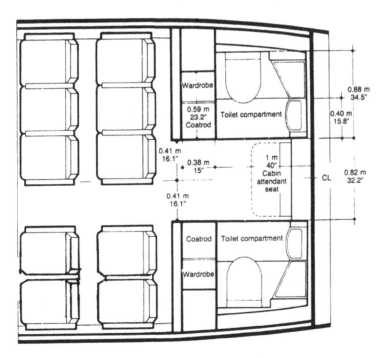

Fig. A9.9 F-100 toilet compartment.

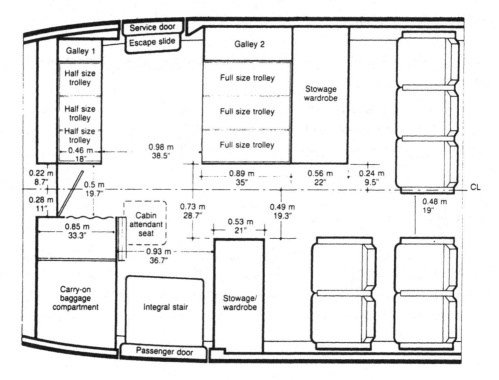

Fig. A9.10 F-100 galley compartments.

abreast airliners cannot use this standard container and manufacturers have developed containers for their aircraft.

Standard main-deck containers are 8 ft × 8 ft cross-section, with lengths of 10 or 20 ft. These are suitable for transport on trucks, railways or ships. Such containers require special modifications to wide-body aircraft, in terms of large cargo doors and strengthened floors.

A9.7 Avionic Components

Table A9.1 shows the navigation and communication equipment list for the Cranfield A-94 long-range airliner project [59] and is typical of the quantities and types of equipment used by long-range aircraft. Avionic equipment components are sized to fit in standard ATR box sizes, the dimensions of which are shown in Table A9.2.

ATR boxes were specified for use on a military aircraft project, as part of the installation shown in Table A9.3 [45], which also shows equipment installed masses.

Basic stowage

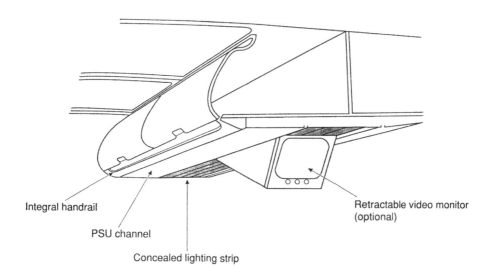

Integral handrail

PSU channel

Concealed lighting strip

Retractable video monitor
(optional)

Optional articulating stowage

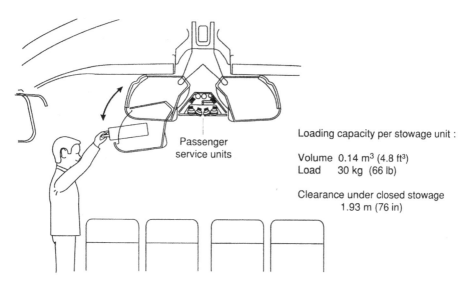

Passenger
service units

Loading capacity per stowage unit :

Volume 0.14 m³ (4.8 ft³)
Load   30 kg (66 lb)

Clearance under closed stowage
1.93 m (76 in)

Fig. A9.11 A340 overhead stowage.

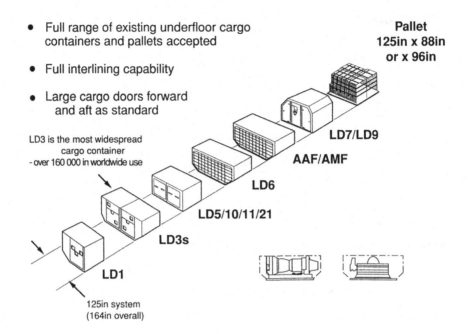

- Full range of existing underfloor cargo containers and pallets accepted

- Full interlining capability

- Large cargo doors forward and aft as standard

Pallet
125in x 88in
or x 96in

LD7/LD9

AAF/AMF

LD6

LD5/10/11/21

LD3 is the most widespread
cargo container
- over 160 000 in worldwide use

LD3s

LD1

125in system
(164in overall)

Type	Base length x width	Max Width	Volume cu ft/m³	Allowable gross lb/kg
LD3	60.4 x 61.5 in 153 x 155 cm	79 in 200 cm	158/4.5	2830/1285
LD1	"	92 in 235 cm	173/4.9	2830/1285
LD5	60.4 x 125 in 153 x 318 cm	-	243/6.9	5660/2570
LD6	"	164 in 417 cm	320/9.1	5660/2570
LD11/21	"	-	243/6.9	5660/2570
Pallet	88 x 125 in 224 x 318 cm	-	379/10.7	8300/3770
LD7	"	-	355/10.0	8300/3770
LD9	"	-	355/10/0	8300/3770

Maximum Height is 64 in (162 cm)

Fig. A9.12 The main types of under-floor containers for wide-body aircraft: full range of existing underfloor cargo containers and pallets accepted; full interlining capability; large cargo doors forward and aft as standard.

Table A9.1   A-94 long-range airliner navigation and communications equipment list

System	ARINC Standard	Number of systems fitted
Air data, inertial reference system (ADIRS)	738	3
Standby instrument	–	1 set
Instrument landing system (ILS)	710	2
Microwave landing system (MLS)	727	–
Very high frequency, omnidirectional radio (VOR) – marker	711	2
Distance measuring equipment (DME)	709	2
Automatic direction finding (ADF)	712	2
Air traffic control transponder (ATC)	718	2
Radio altimeter	707	2
Weather radar	708	1
Ground proximity warning system (GPWS)	723	1
Global positioning by satellite (GPS)	–	1
Very high frequency comms radio (VHF)	–	3
High frequency comms radio (HF)	–	2

A320-family bulk loading

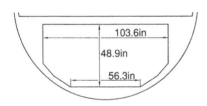

A320-family containerised option

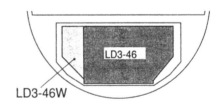

- Forward hold   806ft³/22.82m³
  Aft hold   1 022ft³/28.94m³
  Total hold volume   1 828ft³/51.76m³

- 5 containers forward, 5 aft
  Total of 10 containers + bulk

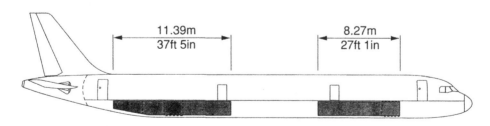

Fig. A9.13 A320 underfloor compartments.

Table A9.2   ARINC standard ATR case sizes

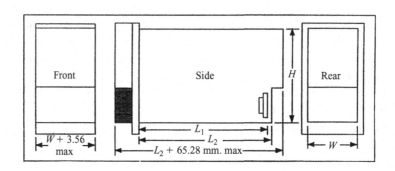

ATR size	Approx. vol (L)	W (±0.76 mm)	L_x (±1.00 mm)	L_2 (max) (mm)	H (max) (mm)
Dwarf	1.56	57.15	318.00	320.50	85.80
¼ short	3.52	57.15	318.00	320.50	193.50
¼ long	5.49	57.15	495.80	498.30	193.50
3/8 short	5.57	90.41	318.00	320.50	193.50
3/8 long	8.69	90.41	495.80	498.30	193.50
½ short	7.70	123.95	318.00	320.50	193.50
½ long	11.88	123.95	495.80	498.30	193.50
¾ short	11.80	190.50	318.00	320.50	193.50
¾ long	18.36	190.50	495.80	498.30	193.50
1 short	15.98	257.05	318.00	320.50	193.50
1 long	24.75	257.05	495.80	498.30	193.50
1½	77.62	390.65	318.00	498.30	193.50

A10 AIRCRAFT WEAPONS

This topic is discussed in Chapter 7, but this appendix gives dimensions and masses for guns, bombs and missiles that are in widespread NATO use, or are likely to be so in the near future.

A10.1 Aircraft Guns

Chapter 7 shows an aircraft installation of an ADEN revolver cannon, a description of the GAU-8 rotary cannon and a performance table for several cannons.

Figure A10.1 gives two views of the M-61 cannon and its specification is:

Table A3.3 Avionic equipment list for an advanced short take-off/vertical landing fighter project

Component	Uninstalled mass (kg)	Size/unit (mm)	Install factor	Installed mass (kg)	Installation notes
Main V/UHF radio	8	½ ATR(S)	1.50	12	Dorsal and ventral antenna
Stand-by V/UHF radio	8	½ ATR(S)	1.50	12	Ventral antenna
IFF transponder	6	3/8 ATR(S)	1.50	9	Ventral antenna
Radio approach aid	8	½ ATR(S)	1.50	12	Ventral antenna
Radio/radar altimeter	4	¼ ATR(S)	1.50	6	Ventral antenna
IRS/GPS	10	180 × 180 × 240	1.50	15	Strapdown, dorsal antenna
Radar	130	As spec.	1.25	162	As spec. (forward)
RWR receiver	2 × 4	90⌀ × 500	1.50	12	180° view forward and aft
RWR processor	15	1 ATR(S)	1.50	22	Internal
Chaff and flare disp.	2 × 14	215 × 270 × 170	1.25	35	Aft facing
2–18 GHz jammer	4 × 15	1 ATR(S)	1.50	90	Internal
FLIR/LRMTS/TI	(3 × 15) +10	1 ATR(S)	1.25	69	As spec. (forward)
Stores management processor	10	¾ ATR(S)	1.50	15	Internal
Stores interface units					
Air-data processors	6 × 3	60 × 150 × 300	1.50	27	Close to stores
Utilities management processor	10	1 ATR(S)	1.50	15	Close to sensors
Display symbol generator	4 × 5	200 × 80 × 300	1.50	30	1 × forward, 1 × rear and 1 × each wing-root
Head-up display	15	260 × 195 × 280	1.26	19	Close to cockpit (for five displays)
CDU/FMS displays	11	As spec.	1.10	12	Instrument panel, centre-top
Data-entry panels	4 × 8	200 × 130 × 310	1.25	40	Instrument panel
FCS processors	2 × 5	150 × 200 × 150	1.25	12	Side consoles
FCS senor packs	4 × 15	1 ATR(S)	1.50	90	Distributed (survivability)
FCS actuator Elect.	4 × 4	150 × 150 × 200	1.50	24	Strapdown central position
FCS batteries	2 × 10	¾ ATR(S)	1.50	30	Near primary control surfaces
	2 × 15		1.25	38	
Total installed masses					
Avionics				562	kg
Instrument panel				64	kg
FCS Electronics				182	kg
Total installed avionics mass				808	kg

Note: IFF = identification, friend or foe; RWR = radar warning receiver; FLIR = forward looking infra-red; LRMTS = laser ranger, marker, target system; CDU/FMS = cockpit display unit/flight management; FCS = flight control system.

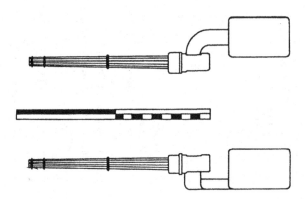

Fig. A10.1 M61 Al Vulcan gun intallation.

Calibre:	20 mm
Masses:	120 kg (gun)
	150 kg (400 rounds)
Overall length:	1880 mm (gun)
No. of barrels:	6
Rate of fire:	4000 rounds/min
Muzzel velocity:	1036 m/s

A10.2 Bombs and Ground Attack Weapons

Figure A 10.2 shows scaled sketches of a range of bombs and ground attack weapons. Their masses and overall dimensions are given, but their width and depths are appropriate to folded-wing, and cruciform wings at 45° to the horizontal. The attack missile is an air-launched cruise missile and the low-altitude dispensers are primarily airfield attack weapons, such as the BL755.

A10.3 Airborne Missiles

There is a vast range of such missiles, but a few of the main NATO weapons will be described:

A10.3.1 AO-120A Amraam
Type: air-to-air semi-active laser homing (Fig. A10.3).

Length:	3650 mm
Body diameter:	178 mm
Wingspan:	630 mm
Range:	50 km

Weapon type	Elevation	Weight kg	Length m	Width* m	Depth* m
AGM-69A Short range attack missile		1000	4.25	0.5	0.5
Apache Low altitude dispenser	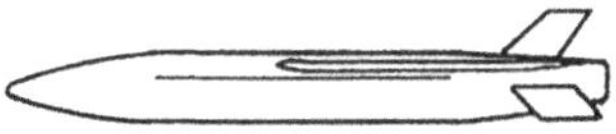	1200	3.85	0.6	0.42
CWS Low altitude dispenser	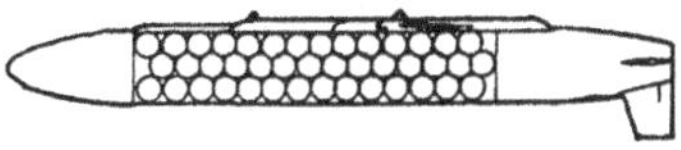	1220	4.15	0.63	0.48
Paveway II (Mk 13/18) 1000 LB Laser- guided bomb	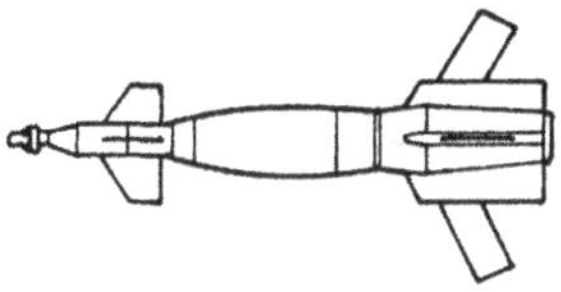	500	3.45	0.52	0.52
BL 755 Cluster bomb	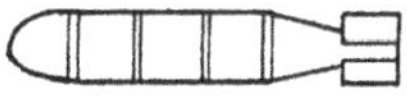	280	2.45	0.4	0.4
Mk 13/18 1000 LB bomb free-fall or retard	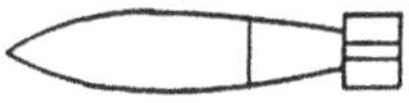	455	2.45	0.42	0.42

Fig. A10.2 Widely used ground attack weapons.

Masses: 157 kg (missile)
                 42 kg (launcher)
Propulsion: solid propellant

A10.3.2 Asraam

Type: air-to-air passive imaging infra-red homing (Fig. A10.4).

Length: 2900 mm
Body diameter: 165 mm
Wingspan: 450 mm
Range: 10 km
Masses: 85 kg (missile)
                 42 kg (launcher)
Propulsion: solid propellant

Fig. A10.3.3 AGM-65 Maverick

Type: air-to-ground semi-active radar, laser, or TV (Fig. A10.5).

Length: 2490 mm
Body diameter: 305 mm

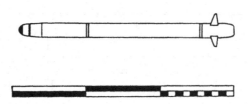

Fig. A10.3 Aim 120A AMRAAM.

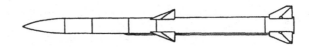

Fig. A10.4 ASRAAM.

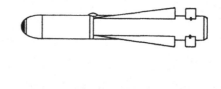

Fig. A10.5 AGM-65 Maverick.

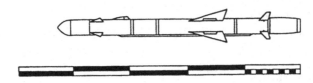

Fig. A10.6 Alarm.

Wingspan: 720 mm
Range: 25 km (F/G)
Masses: 307 kg (missile, F/G)
                 37 kg (launcher)
Propulsion: Solid propellant

Fig. A10.3.4 Alarm
Type: anti-radiation missile (Fig. A10.6).

Length: 4300 mm approx.
Body diameter: 230 mm approx.
Masses: 260 kg approx. (missile)
                 70 kg approx. (launcher)

APPENDIX B: A-90 PARAMETRIC STUDY. EXAMPLE: THE A-90 500-SEAT AIRLINER

B1 INTRODUCTION

A parametric study is a vital initial stage in the conceptual design of an aircraft. The example in this appendix has been extracted from Fielding [29] and formed the basis of the A-90 group design project. An introduction to this project has been given in Chapter 10 of this book. The main points of the design requirements are listed in Section 10.5.1.

The parametric study was based on the method by Loftlin [60]. This was an initial attempt to determine basic aircraft parameters such as wing loading, thrust/weight ratio, wing aspect ratio, take-off and landing lift coefficients, and cruise lift/drag ratios. This method relies heavily on empirical data and cannot be safely used for aircraft which are not of the conventional subsonic transport type. The method uses knots, lb, and feet units, which were converted to SI units at the end of the study. The method starts with a large design space and then excludes designs that cannot satisfy landing and take-off distances, second segment climb or missed-approach requirements. It then goes on to examine cruise performance. An extra requirement that came from the aircraft specification was the 15 000 ft ceiling after a single engine failure.

B2 LANDING FIELD DISTANCE

Figure B2.1 shows an empirical plot of the relationship between approach speed and FAR landing field length. The required value of 5650 ft gave, $V_A^2 = 19\,100$ therefore = 138 Knots.

Having this value, Fig. B2.2 was used to determine the wing-loading parameter. This process required the estimation of the approach lift coefficient. The variable camber flaps required by the specification could be expected to produce a landing $C_{L_{max}}$ of 2.5. The approach lift coefficient was found from:

$$C_{L_A} - C_{L_{max}}/1.3^2 = 1.48.$$

This follows because stall speed, V_S occurs at $C_{L_{max}}$ and approach speed, $V_A = 1.3V_s$. This value and the approach speed on Fig. B2.2 gave the landing parameter

$$\sqrt{\frac{W_L/S}{\delta}} = 9.95,$$

at ISA, $\delta = 1$, therefore $W_L / S = 99$ lb/ft^2.

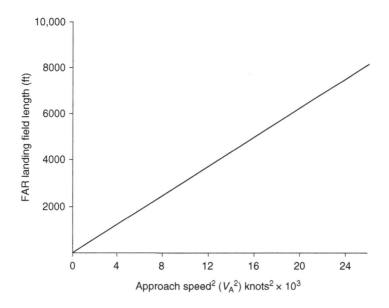

Fig. B2.1 Landing field length.

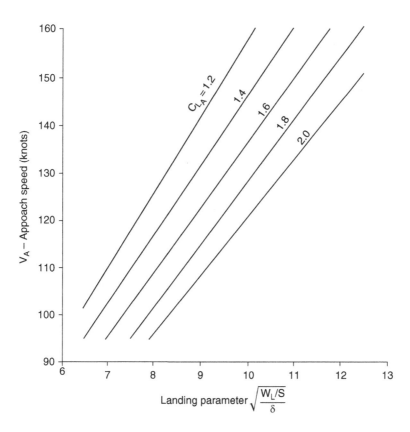

Fig. B2.2 Approach speed.

The parametric study was carried out in terms of gross-weight, W_G, so a relationship between that and the landing weight (W_L) had to be determined. Study of comparable-range aircraft indicated that

$$\frac{W_L}{W_G} = 0.83,$$

this then gave the value to satisfy landing requirements as

$$\frac{W_G}{S} = \frac{W_L}{S}\frac{1}{0.83} = 119.3\,\text{lb/ft}^2.$$

This is the landing boundary shown on Fig. B8.1.

B3 TAKE-OFF FIELD LENGTH

Using Fig. B3.1 with a length of 8300 ft gave $\dfrac{W_G S}{\delta C_{LT}(T_0/W_G)} = 220.$

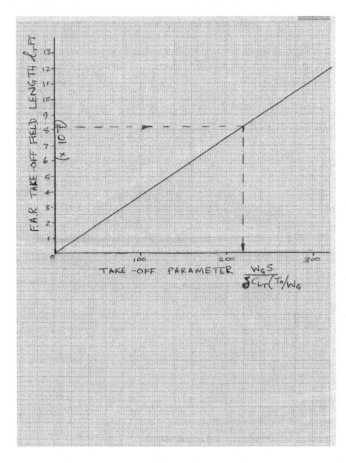

Fig. B3.1 Take-off field length.

Assuming $C_{L_T} = 1.8$ (max. lift coefficient in take-off configuration)

$$\delta = 1(\text{ISA}, \text{SL})$$

$$\frac{W_G/S}{T_0/W_G} = 220 \times 1.8 = 396.$$

Using this value, a suitable range of wing loadings gave:

W_G/S (lbf/ft^2)	90	100	110	120	130
T_0/W_G	0.227	0.252	0.28	0.303	0.328

These were plotted as the take-off boundary in Fig. B8.1.

B4 SECOND SEGMENT CLIMB

This is the part of the take-off between 35 and 400 ft altitude at speed V_2. Airworthiness requirements state that in the event of an engine failure during this period, a twin-engined aircraft must sustain a 2.4% climb with flaps in take-off setting and undercarriage retracted. Two engines were chosen as the most economic solution, consistent with safety.

We need to find C_{L_2}, which is the lift coefficient associated with V_2. As V_2 is 1.2 times the stalling speed,

$$C_{L_2} = 1.8/1.44 = 1.25$$

Using Fig. B4.1 for twin-engined aircraft with this value of C_{L_2} we obtain the following T_0/W_G ratios for a range of aspect ratios:

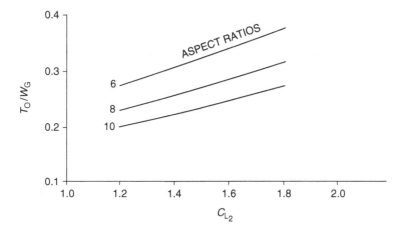

Fig. B4.1 Second segment climb – two engined aircraft.

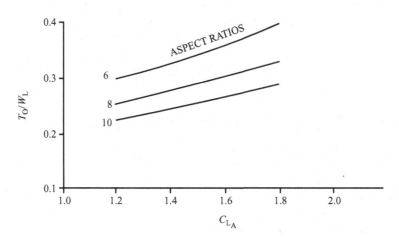

Fig. B5.1 Missed approach two-engined aircraft.

Aspect ratio	6	8	10
T_0/W_G	0.29	0.24	0.21

These values are included as second segment boundaries in Fig. B8.1.

B5 MISSED APPROACH

This must be considered in relation to the landing manoeuvre. This occurs when the aircraft is on final landing approach, but does not land for any reason. Power is applied and the aircraft circles, usually to try another landing. Airworthiness authorities require the installation of sufficient thrust to enable adequate climb gradients with the aircraft in approach configuration. This would include approach slats and flaps and undercarriage. Figure B5.1 shows empirical figures of required T_0/W_G for various aspect ratios and engine requirements.

With $C_{L_A} = 1.48$ from Section B2 we obtain:

Aspect ratio	6	8	10
T_0/W_L	0.34	0.29	0.25
T_0/W_G (as $W_L/W_G = 0.83$)	0.28	0.24	0.21

These values are included in Fig. B8.1 as missed approach boundaries.

B6 CRUISE PERFORMANCE

The optimization of cruise performances is a complex operation in which engine performance and aircraft lift/drag ratios are matched. Figure B6.1 gives a chart to obtain an initial estimate of the maximum lift/drag ratio. We need to know the parameters $\pi d^2/4S$ and $1d$.

Initial fuselage layout work, assuming a multiple circular arc cross-section with main deck ten-abreast seating yielded $d = 23$ ft and $l/d = 8.3$ from initial fuselage layout drawings.

An initial wing area of 4200 ft^2 was assumed (based on the Boeing 777). Using Fig. B6.1 with the above figures gave:

Aspect ratio	$L/D_{\max}$	C_{L_m}	$C_{D_{0_{\text{ref}}}}$
7	16.7	0.55	0.017
9	18.9	0.64	0.017
10	20	0.67	0.017

Figure B6.1 was based on a reference aircraft flying at a given Reynolds number very similar to the study aircraft, and was therefore valid. (Note: Reynolds number correction [60] will have to be used for aircraft with different Reynold's numbers.)

It was thought that flying at $L/D_{\max}$ would be optimistic and have a high cruise C_L, therefore a value of 0.97 $L/D_{\max}$ would be used. Using this value in Fig. B6.2 gave a $\overline{C}_L$ value of 0.78. Applying this factor to the C_{L_m} above, the following values were produced:

Aspect ratio	L/D	$C_{L_{\max}}$ (cruise)
7	16.2	0.429
9	18.3	0.499
10	19.4	0.52

These values were associated with different combinations of cruise altitude and wing loading, as shown in Fig. B6.3. They were checked for a cruise Mach number of 0.82 (economical) (Table B6.1).

The next step was to check the thrust/weight ratios associated with these wing loadings. Section B3 gave

$$T_0/W_G = \frac{1}{(T_C/T_0)(L/D)_{\max}},$$

where T_C/T_0 is a measure of thrust decay with altitude and Mach number. Figure B6.4 shows generalized curves for a modern high compression-ratio turbo-fan with a bypass ratio of 4.5, which

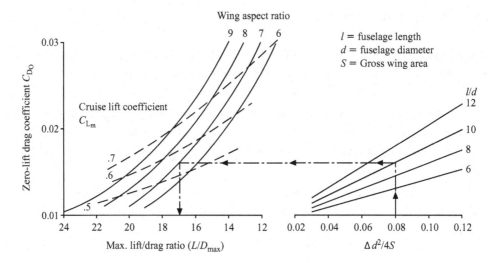

Fig. B6.1 Lift–drag.

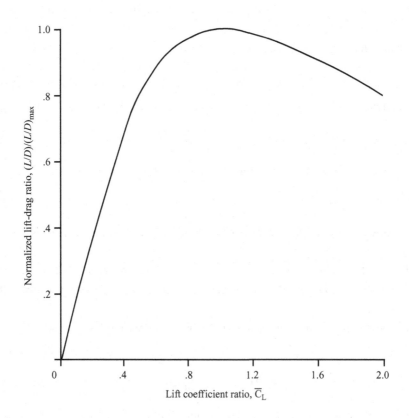

Fig. B6.2 Normalized *L/D*.

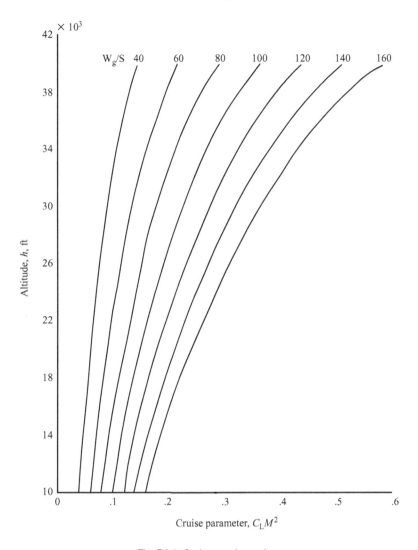

Fig. B6.3 Cruise mach number.

should be a suitable class of engine for the A-90. Using Fig. B6.4 we obtained the values of T_C/T_0 and therefore the T_O/W_G given in Table B6.2, which ae plotted as cruise boundaries on Fig. B8.1.

   The specification called for a cruise altitude of 'at least 39 000 ft'. Cruise altitudes of 39 000, 40 000 and 41 000 ft were cross-plotted from the cruise curves and are drawn on Fig. B8.1.

B7 CEILING WITH ONE ENGINE INOPERATIVE

The parameters on Fig. B8.1 were pointing towards a wing loading of about 120 lb/ft², but no clear aspect ratio had emerged. It was decided to check the average value of 9 and a reduced Mach number of 0.65 for this requirement.

Table B6.1   Cruise conditions

Aspect ratio	L/D	C_L	$C_L M^2$	$W_G/S(\text{lb/ft}^2)$	$h(\times 10^3 \text{ ft})$
7	16.2	0.429	0.288	90	37.5
				100	33
				110	31
				120	29
9	18.3	0.499	0.336	90	41
				100	36.5
				110	35
				120	32.6
10	19.4	0.52	0.35	90	41.5
				100	38
				110	36
				120	34

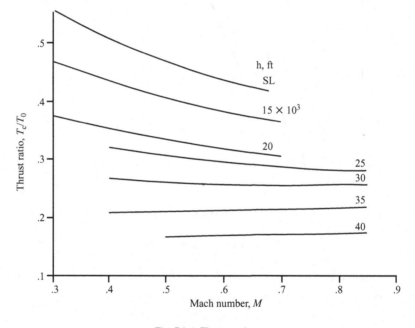

Fig. B6.4 Thrust ratios.

This produced a lift coefficient of 0.355, which gave the lift coefficient ratio

$$\frac{0.355}{0.64} = 0.56.$$

(Figure B6.2 then gave a normalized lift/drag ratio of 0.86.) Therefore at 15 000 ft, $L/D = (L/D_{max})$
$\times 0.86,$

Table B6.2

Aspect ratio	$W_G/S(\text{lb/ft}^2)$	T_c/T_o	T_o/W_G
7	90	0.195	0.316
	100	0.235	0.26
	110	0.25	0.247
	120	0.26	0.237
9	90	0.16	0.341
	100	0.205	0.267
	110	0.215	0.25
	120	0.23	0.238
10	90	0.145	0.355
	100	0.185	0.279
	110	0.205	0.251
	120	0.22	0.234

$$T_o/W_G = \frac{1}{(T_C/T_0)(L/D)} = 16.25,$$

where $T_C/T_0 = 0.4$ from modern engine (max continuous) $= 1/(0.4 \times 16.25) = 0.154$.

This is based on both engines working, therefore the single engine case = 0.308. A similar check for a wing with an aspect ratio of 10 gave a value of 0.284. This was the final parameter inserted into Fig. B8.1. This figure does not include the effect of windmilling or yaw drag, which will have to be checked later.

B8 ARRIVAL AT THE MATCH POINT

Figure B8.1 shows the design space available when all six performance parameters are satisfied. For a good, lightweight design we needed to maximize wing loading and minimize thrust loading, indeed, to be as close as possible to the bottom right-hand corner of the graph.

The main intersection is that of the take-off and landing boundaries at $W/S = 119.3$ lb/ft^2 and $T_c/W_G = 0.301$. This is more than adequate for both the second segment climb and missed approach for all of the considered aspect ratios. There seems to be more than enough thrust for cruise, but this will leave a margin for possibly increased thrust lapse with high-bypass engines. The maximum cruise altitude is adequate, at between 40 000 and 41000 ft.

The deciding factor on wing aspect ratio was the single engine failure ceiling. The minimum suitable aspect ratio was 9, which gave a T_o/W_G of 0.308, which was chosen as the final value.

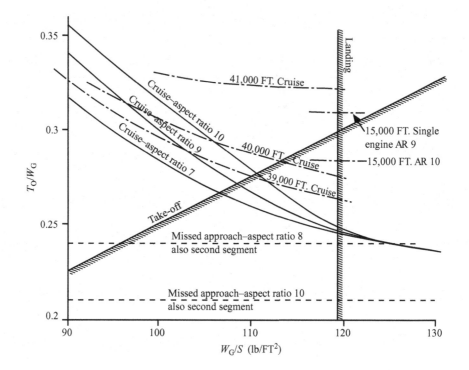

Fig. B8.1 Summary of results.

Summarizing the match point and its implicit parameters:

Max. wing loading	W_G/S	= 119.3 lb/ft$_2$	= 583 kg/m$_2$
Max. thrust loading	T_o/W_G	= 0.308	
Wing aspect ratio	AR	= 9.0	
$C_{L_{max}}$ take-off		= 1.8	
$C_{L_{max}}$ landing		= 2.5	

W_L/W_G	= 0.83
Opt. cruise C_L	= 0.5
L/D cruise	= 18.3
Max. cruise altitude	= 40 500 ft = 12.34 km
$M_{economical}$	= 0.82
M_c	= 0.86
M_D	= 0.92

APPENDIX C: THE PREDICTION OF AIRCRAFT RELIABILITY AND MAINTAINABILITY TARGETS

C1 INTRODUCTION

The author and V. C. Serghides have performed statistical analyses to produce combat aircraft reliability and maintainability prediction methods [61]. These are applicable for use at the conceptual design stage, because they only require the use of readily available parameters such as wing span, engine thrust, mass, etc. These methods predict whole-aircraft values of confirmed defects per 1000 flying hours and defect maintenance hours per 100 flying hours. Predictions are also made for individual systems and allowances may be made for technology improvements, relative to the empirical database used in the method derivation. These may be used as targets for reliability performance of individual systems during the preliminary design stage.

Earlier work by the author produced a similar method for the prediction of commercial aircraft dispatch reliability [62]. The whole-aircraft equations produced are reproduced in Section C2, below.

C2 COMMERCIAL AIRCRAFT DISPATCH RELIABILITY PREDICTION

The work reported in reference 62 showed that some systems exhibited different traits according to the type of airline operation. For example long-haul aircraft tended to have higher delay rates because they may be away from their home base for more than a week, and defects might accumulate, whereas short-haul aircraft tended to return to their home bases more often. Other considerations were also affected by the operational type, in such things as ATA Chapter 28, fuel systems, where long-haul aircraft had more complicated systems that were more delay-prone than those of short-haul aircraft. To cater for these effects, separate models were made for the relevant systems.

All the aircraft in the sample had turbo-fan engines of some sort. Some were early types with bypass ratios of less than unity and some were of the more modern 'big-fan' type with bypass ratios on the order of five. These two types of engines, again, exhibited different traits, so separate curves were drawn for them. After the graphs were plotted, it was decided that, for most systems, a linear regression analysis would give the fairest result. This process was carried out by the use of a least-squares fit program.

After this process the results of all the system analyses were added together to produce the following formulae for the total aircraft:

Table C2.1

	Short-haul, small fan engines	Short haul, big fan engines	Long haul, small fans	Long haul, big fans
C_1	0.2777	−0.4451	−0.7357	−0.0129
C_2	0.154	0.235	0.154	0.235
C_3	1.84	1.84	2.175	2.175
C_4	−0.0038	−0.0038	−0.0102	−0.0102
C_5	0.0419	0.0419	0.0419	0.0419
C_6	0.7348	1.1426	0.7348	1.1426

C2.1 Short-Haul Aircraft

$$
\begin{aligned}
\text{Delay rate} = {} & C_1 + \left(C_2 \times W_T \times 10^{-6}\right) + \left(C_3 \times \frac{W_T}{T} \times 10^{-6}\right) + \left(C_4 \times \frac{P}{100 \times T}\right) \\
& + \left(\frac{C_5 \times P}{100}\right) + \left(10^{\left(3.6 \times (W_T/T) \times 10^{-6}\right) - 1.477}\right) \\
& + \left(0.3075 \log_{10} V^2_{AT} \times W_L \times 10^{-8}\right) \\
& + \left(\frac{SLST \times C_6}{100\,000}\right)
\end{aligned}
\tag{C2.1}
$$

C2.2 Long-Haul Aircraft

This equation is similar to Equation (C2.1) except that the sixth term is

$$
10^{\left(2.29 \times (W_T/T) \times 10^{-6}\right) - 0.887}
\tag{C2.2}
$$

where W_T is the aircraft maximum take-off mass (lb); T is the average flight time (h); P is the maximum number of passengers; V_{AT} is the aircraft approach speed (ft/sec); W_L is the aircraft maximum landing mass (lb); $SLST$ is the aircraft total sea level static thrust (lb).

The terms delay rate and description of the ATA systems are shown in Chapter 10. Note that in both formulae, term 6 caters for ATA Chapter 27, flying controls and term 7 caters for ATA Chapter 32, landing gear. The coefficients are shown in Table C2.1 and the results for individual systems are shown in Table C2.2.

The service entry dates of aircraft types in the study sample cover a period of about 20 years. Whilst there has been little increase in cruise speed over this period, there have been considerable improvements in technology, particularly in electronic components. Later aircraft also

Table C2.2   Individual formula results

ATA	System	Delay rate prediction Equation
21	Air conditioning	DR = 0.0006(CS) − 0.2132
22	Auto pilot	DR = −0.0063(FL) + 0.0318
23	Communications	DR = −0.00018(NP) + 0.0449
24	Electric power	DR = −7E-07(MTW/FL) + 0.0707
25	Equipments	DR = 0.0005(CS) − 0.1904
26	Fire protection	DR = −0.0001(Thr) + 0.0528
27	Flight control	DR = −0.0007(NP/FL) + 0.1879
28	Fuel	LnDR = −0.0166(NP/FL) − 2.1379
29	Hydraulic power	DR = −5E-07(MTW) + 0.1428
30	Ice protection	DR = 0.0007(CS) − 0.2627
31	Instrument	DR = −1E-06(MTW/FL) + 0.0696
32	Landing gear	DR = −0.0008(NP) + 0.2704
33	Light	DR = 0.0074(FL) + 0.0039
34	Navigation	DR = −0.0003(Thr) + 0.1419
35	Oxygen	DR = −0.000094(NP) + 0.0234
36	Pneumatic	DR = −0.0006(CS) + 0.323
38	Water waste	DR = 3E-07(MTW) − 0.0088
49	APU	DR = −0.014(FL) + 0.061
51	Structure	DR = −6E-07(MTW) + 0.0777
52	Doors	DR = −0.0005(NP) + 0.1239
71–80	Powerplant	DR = 0.0014(CS) − 0.4259

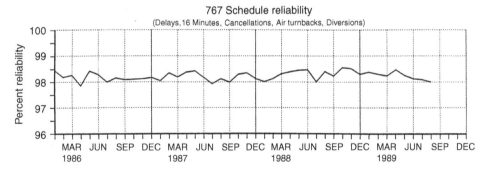

Fig. C2.1 Boeing 767 schedule reliability (delays >15 min, cancellations, air turnbacks, diversions).

Table C2.3

ATA	System	Delay rate prediction Equation
21	Air conditioning	$DR = 0.0009(CS) - 0.4041$
22	Auto pilot	$DR = 0.0004NP/FT - 0.0127$
23	Communications	$DR = 5E{-}07MTW - 0.065$
24	Electric power	$DR = 1E{-}06MTW/FT + 0.0323$
25	Equipments	$DR = 0.0016NP/FT - 0.0475$
26	Fire protection	$DR = 0.0003NP - 0.0379$
27	Flight control	$DR = 0.0029NP/FT - 0.0772$
28	Fuel	$DR = 0.0005NP - 0.0549$
29	Hydraulic power	$DR = 0.0004NP - 0.0107$
30	Ice protection	$DR = 2E{-}06MTW/FT - 0.0306$
31	Instrument	$DR = 2E{-}07MTW/FT + 0.0095$
32	Landing gear	$DR = 0.0007NP - 0.0291$
33	Light	$DR = 1E{-}06MTW/FT - 0.0144$
34	Navigation	$DR = 0.0033CS - 1.4863$
35	Oxygen	$DR = 9E{-}05NP/FT - 0.0027$
36	Pneumatic	$DR = -0.0108FT + 0.1855$
38	Water waste	$DR = 0.0011CS - 0.5014$
49	APU	$DR = 0.0009NP/FT - 0.0028$
51	Structure	$DR = 0.0015CS - 0.6801$
52	Doors	$DR = 0.0005NP - 0.0878$
71–80	Powerplant	$DR = -0.0003Thr + 0.4981$

Table C2.4

CS (Kts)	Cruise Speed
MTW (Kg)	Maximum Take Off Weight
MTW/FL (Kg/hr)	Maximum Take Off Weight / Flight Length
NP/FL (hr^{-1})	Number Of Passengers / Flight Length
NP	Number Of Passengers
FL (hr)	Flight Length
Thr (kN)	Aircraft Thrust

benefitted from more rigorous reliability analyses and quality control. It was decided to make some allowance for this effect on the data used in the formulae derivation.

The advent of the 'big-fan' engines made a significant change in reliability, so it was decided to make corrections for both 'small-fan' and 'big-fan' engines.

Independent fleet delay rate figures were obtained and compared with the results of the above equations. Formula percent errors were then calculated and correction factors determined. Initial work suggested that results obtained from the above formulae should be factored to reflect a 1% improvement in reliability for every year of the service-entry date beyond 1969. This is only applicable for the first model of an aircraft type. For example, the formulae were applied to a Boeing 767 and predicted a delay rate of 2.09%. The technology improvement factor of 0.9 gave a corrected value of 1.88%. This equates to a dispatch reliability of 98.12%, which may be compared with the mature value for the aircraft, shown in Fig. C2.1 The term schedule reliability is used in the figure, but is identical to dispatch reliability.

More recent work was performed under the author's supervision [63]. This had a similar methodology to that above, but utilized data from more modern aircraft types. The method enabled the prediction of delay rates for each individual system, and then summed them for whole aircraft prediction. Table C2.2 shows the equations for short-haul aircraft, and Table C2.3 for long-haul. Table C2.4 shows the aircraft parameters used in the previous two tables.

REFERENCES

1. HABERLAND, C., THORBECK, J. and FENSKE, W. A computer augmented procedure for commercial aircraft preliminary design and optimization. *Proceedings of the 14th Conference of the International Council of Aeronautical Sciences*, Paper ICA84-4.8.1, 943–53, 1984.

2. ANON. *Current Market Outlook*. Boeing Commercial Airplane Group, 2015.

3. ANON. Global Market Forecast. Airbus 2015.

4. GREFF, E. Aerodynamics design and integration of a variable camber wing for a new generation long/ medium range aircraft. *Proceedings of the International Council of Aeronautical Sciences Congress, Jerusalem, Israel*, Paper ICAS 88-2.2.3. 1988.

5. FIELDING, J. P. *500-Passenger Short-Range Airliner Project Specification. DES 9000.* Cranfield Institute of Technology, 1991.

6. ANON. *Jane's All the World's Aircraft*. Jane's Information Group, Coulsden, Surrey, UK. Published annually.

7. SMITH, H. *AVD 9900 Uninhabited Tactical Aircraft U-99*. Project specification. Cranfield University, 1999.

8. HOWE, D. *Introduction to the Basic Technology of Stealth Aircraft*. ASME Paper 90-GT-116. American Society of Mechanical Engineers. 1990.

9. HODGSON, S. W. *Preliminary Design Study of a Stealth Configured Tactical Bomber*. MSc thesis, Cranfield Institute of Technology, 1987.

10. JONKERS, R. P. *Conceptual Design of an Executive Filtrator Aircraft*. MSc thesis, Cranfield Institute of Technology, 1991.

11. TORENBEEK, E. *Synthesis of Subsonic Airplane Design*. Delft University Press. 1982.

12. CONWAY, H. G. *Landing Gear Design*. Chapman and Hall, London, 1958.

13. CURREY, N. S. *Landing Gear Design: Principles and Practices*. American Institute of Aeronautics and Astronautics 1988.

14. JANG, S. S. *Prediction of Manufacturing Costs and Factory Requirements for the Production of the Metal T-84 Basic Training aircraft.* MSc thesis. Cranfield Institute of Technology, 1985.

15. PUGH. P. G. Working top-down: cost estimating before development begins. *Proceedings of the Institute of Mechanical Engineers.* 206, 1993.

16. LEVENSON, G. S., BOREN, H. E., TIHANSKY, D. P., Jr. and TIMSON, F. *Cost Estimating Relationships for Aircraft Airframes.* Rand Report R-761-PR. The Rand Corporation, Santa Monica, California, USA. 1972.

17. DOGANIS, R. *Flying Off Course: The Economics of International Airlines, 5th Edition.* Taylor and Francis. 2010.

18. ANON. *Short-Medium Range Aircraft: AEA Requirements, G (T) 5656.* Association of European Airlines, 1989.

19. FIELDING, J. P. *AVD 1201 Introduction to Aircraft Reliability Design Process and Targets.* Cranfield University, 2012.

20. PECK, M. Pentagon unhappy about drone aircraft reliability. *National Defense Magazine*, May 4th, 2009.

21. ANON. Grim Reaper Rate. *Aviation Week and Space Technology Magazine*, May 2003.

22. ANON. EMBRAER: Raising the Bar. *Aerospace Technology Magazine, Royal Aeronautical Society.* May, 2013.

23. ANON. *MSG-2 Airlines/Manufacturer Maintenance Program Planning Document.* Air Transport Association, Washington DC, USA. 1970.

24. FIELDING, J. P. *AVD 0512 Introduction to Maintainability.* Cranfield University, 2013.

25. SERGHIDES, V. C. *Design Synthesis for Canard Delta Combat Aircraft.* PhD thesis. Cranfield Institute of Technology, 1988.

26. SERGHIDES, V. C. and FIELDING, J. P. A reliability and maintainability prediction method for aircraft conceptual design. *Proceedings of the 16th Congress of the International Council of the Aeronautical Sciences, Jerusalem, Israel*, 1988.

27. FIELDING, J. P. and VAZIRY-Z. M. A. F. Avionics reliability enhancement modelling for aircraft conceptual design. AIAA 95 - 3906. *Proceedings 1st AIAA Aircraft Engineering Technology and Operations Congress, Los Angeles CA, USA*, September, 1995.

28. ANON. Advertisement for Hawker 1000. *Flight International Magazine*, September, 1993.

29. FIELDING, J. P. *Report of the Project Design of the Cranfield A-90 Short Haul 500-seat Airliner Project*. CofA Report NFP 9103. Cranfield Institute of Technology, 1991.

30. FIELDING, J. P. 500-seat short range airliner project. *Proceedings Aerotech 94 Conference Institute of Mechanical Engineers, Birmingham UK*, January, 1994.

31. ROSKAM, J. *Airplane Design. Part VIII. Airplane Cost Estimation Design, Development, Manufacturing and Operating*. The University of Kansas. 1990.

32. LIM, P. T. L. *Technical and Commercial Viability of the A-90 500 Seat Short Haul Airliner*. MSc thesis, Cranfield Institute of Technology. September, 1992.

33. FIELDING, J. P. High capacity subsonic transport projects. *Proceedings of the International Congress of the Aeronautical Sciences. Sorrento, Italy*, Paper ICAS-96-3113, September, 1996.

34. WARD, D. *Project Cancelled, 2nd edn*. Tri-Service Press Ltd. 1990.

35. BARNETT-JONES, F. *TSR-2-Phoenix or Folly?* GMS Enterprises. 1994.

36. WILSON, B. Lecture Notes: Lessons Learned - Part 1 - Northrop F-20 Fighter. Lectures given at Cranfield University, 21st May, 1995.

37. CHICKEN, S. H. *Conceptual Design Studies for Amphibious Aircraft*. PhD thesis. Cranfield University, 1998.

38. FIELDING, J. P. *Supersonic Vertical Take-Off and Landing Aircraft - S-83*. DES 8300. Cranfield Institute of Technology, July, 1984.

39. SMITH, H. *DES 1100 Advanced Blended Wing Body High Capacity Airliner, BW-11*. Project Specification. Cranfield University, September, 2011.

40. MISTRY, S. SMITH, H. and FIELDING, J.P. Commercial aircraft design for reduced noise and environmental impact. *Variational Analysis and Aerospace Engineering*. Springer Science and Business. 2009.

41. FIELDING, J.P. STOCKING, P. and SMITH, H. Comparison of fuel burn and noise characteristics of novel aircraft configurations. *International Congress of the Aeronautical Sciences. ICAS 2010. Nice, France*, September, 2010.

42. ROEDER, J. The case for an ultra high capacity container aircraft. *GSE Today Magazine*. July, 1996.

43. CORBAS, N. *Conceptual Design of a Large Wing in Grand Effect Transport Aircraft*. MSc thesis, Cranfield University, 1996.

44. FIELDING, J.P., LAWSON, C.P., MARTINS-PIRES, R.M. and MONTERZINO, G. A. I. Design, build and flight test of the DEMON demonstrator UAV. *11th AIAA Aviation Technology, Integration and Operation (ATIO) conference. AIAA, Virginia Beach, VA, USA*. September, 2011.

45. HEWSON, R. *A Project Design Study of an A/STOVL Strike Fighter*, MSc research thesis. Cranfield Institue of Technology, 1993.

46. FIELDING, J. P. and SMITH, H. *Advanced Training Aircraft : T-91 Project Specification*. DAeT 9101, Cranfield Institute of Technology, 1992.

47. FIELDING, J. P. and JONES, R. I. *Tactical Fighter: TF-89 Project Specification*. DES 8900, Cranfield Institute of Technology, 1990.

48. FIELDING, J. P. Project design of alternative versions of the SL-86 two-stage horisontal take-off space launcher. *Proceeedingsof the International Council of Aeronautical Sciences (ICAS), Stockholm, Sweden*, 1990.

49. VAN DEN BERG, R. *A-90 Powerplant Installation*, MSc thesis. Cranfield Institute of Technology, 1991.

50. ANON. *AIAA Aerospace Design Engineers Guide*, American Institute of Aeronautics and Astronautics, 1987.

51. ABBOTT, I. and VON DOENHOFF, A. *Theory of Wing Sections*, Dover Publications, 1959.

52. FULKER, J. L. *Aerodynamic Data for Three Supercritical Aerofoils*, RAE Reports and Memoranda R & M 3820, Her Majesty's Stationary Office, UK, 1975.

53. ASHILL, P. R., WOOD, R. F. and WEEKS, D. J. *An Improved Semi-inverse Version of the Viscous Garabedian and Korn Method (VGK)*, RAE TR 87002, Her Majesty's Stationary Office, UK, 1987.

54. ANON. NASA SC(3)-0712(B) Supercritical Airfoil Section, from NASA TM-86371, NASA, USA, 1986.

55. ANON, Douglas Santa Monica Airfoil Section DSMA 523, from NASA TM-81336, NASA, USA, 1981.

56. McLUNDIE, W. M. *Main Landing Gear Design for the A-94 Red Ultra High Capacity Airliner*, MSc thesis. Cranfield University, 1995.

57. ANON. *Range List*, Dunlop Ltd, Aircraft Tyres Division, Birmingham, UK, 1992.

58. ANON. *Design and Airworthiness Requirements for Services Aircraft*, Defence Standard 00-970, Ministry of Defence, UK, 1983.

59. CHUDOBA, B. *A-94 Ultra High Capacity Airliner, Reliability and Maintainability Design*, MSc thesis. Cranfield University, 1995.

60. LOFTIN, L. K. JR, *Subsonic Aircraft: Evaluation and Matching of Size to Performance*. NASA Ref. Publication 1060, August, 1980.

61. SERGHIDES, V. C. *Development of a Reliability and Maintainability Prediction Methodology for the Aircraft Conceptual Design Process*. MSc thesis, Cranfield University, 1995.

62. FIELDING, J. P. A Transport Aircraft Dispatch Reliability Formula. *Proceedings of the 2nd National Reliability Conference, Birmingham, UK*. UKAEA, 1979.

63. BIN EID, M. *Aircraft Systems Design and Dispatch Reliability Prediction*, PhD thesis. Cranfield University, 2005.

64. ANON. *Flightpath 2050: Europe's Vision for Aviation*. European Commission, European Union, 2011.

65. SMITH, H *Liquefied natural gas ultra-green airliner LNG-14*. Project Specification. Cranfield University, March, 2014.

INDEX

Printed in the United States
by Baker & Taylor Publisher Services